Karsten Kremer

Invariants of complex and p-adic origami-curves

Invariants of complex and p-adic origami-curves

by
Karsten Kremer

KIT Scientific Publishing

Mathematics Subject Classification (2010): 14H30, 32G15, 14G22, 20E08

Dissertation, Universität Karlsruhe (TH)
Fakultät für Mathematik,
Tag der mündlichen Prüfung: 22.07.2009
Referent: Prof. Dr. Frank Herrlich
Korreferent: PD Dr. Stefan Kühnlein

Impressum

Karlsruher Institut für Technologie (KIT)
KIT Scientific Publishing
Straße am Forum 2
D-76131 Karlsruhe
www.uvka.de

KIT – Universität des Landes Baden-Württemberg und nationales
Forschungszentrum in der Helmholtz-Gemeinschaft

KIT Scientific Publishing 2010
Print on Demand

ISBN 978-3-86644-482-9

Preface

Take finitely many complex unit squares, glue the right edge of every square to the left edge of any square, and every upper edge to a lower edge, such that an orientable compact Riemann surface arises. For example we can use the following four squares

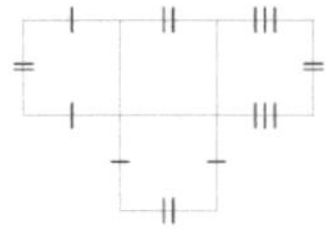

and glue edges with equal marks together. We call such a surface an *origami*.

By mapping each square to the complex unit square glued with itself we get a natural covering of a torus. This covering is ramified only at the vertices of the squares, which are all mapped to the same point on the torus, hence we have a covering with only one branch point. On the other hand given a covering of the torus $\mathbb{C}/(\mathbb{Z}+i\mathbb{Z})$ by a Riemann surface X which is ramified at most over 0 we can lift the unit square to X and hence get a description of X by glued squares.

If we glue parallelograms instead of squares we get a family of Riemann surfaces for any given glueing rule. In the moduli space $\mathcal{M}_g$ of Riemann surfaces of genus g this family forms a one-dimensional subset, on which there is an action of $\mathrm{SL}_2(\mathbb{R})$ corresponding to the stretching and shearing of the parallelograms. The interesting point about origamis is that this subset always is an algebraic curve ([Loc], Prop. 3.2 ii)). We call such a curve in moduli space the corresponding *origami-curve*.

Origamis are a special case of *translation surfaces*, which are Riemann surfaces with an atlas where (almost) every coordinate change map is a translation. As above there is an $\mathrm{SL}_2(\mathbb{R})$ action on those translation surfaces, defined by stretching and shearing the coordinate charts. The orbits lead to isometric embeddings of the upper half plane $\mathbb{H} \cong \mathrm{SL}_2(\mathbb{R})/\mathrm{SO}_2(\mathbb{R})$ into Teichmüller space, whose image is called *Teichmüller disk*. If the projection of this disk in the moduli space $\mathcal{M}_g$ is an algebraic curve (as is the case for origamis), then the translation surface is called a *Veech surface*. These surfaces and especially origamis were first studied by Veech [Vee] and Thurston [Thu].

Since origamis (also known as *square-tiled surfaces*) are dense in $\mathcal{M}_g$ ([HS3], §1.5.2) they offer a good opportunity to study the moduli space $\mathcal{M}_g$. We are going to give an overview on origamis and origami-curves in the first two chapters. They have been studied recently for example by Lochak [Loc],

Zorich [Zor], Schmithüsen [Sch2] and Herrlich [HS1]. We will also present an algorithm to find a set of origamis representing every origami-curve with given degree or Galois group.

In the third chapter we will study some invariants which can be used to distinguish different origami-curves. Recent papers on this topic include [HL] and [Mc2]. In Appendix A.2 we calculate some of these invariants for all origami-curves of origamis made up of up to eight squares.

By GAGA there is a one-to-one correspondence between Riemann surfaces and complex nonsingular projective curves. Thus an origami is a covering of a nonsingular projective curve over an elliptic curve which may be ramified only over 0. This definition can be generalized to other ground fields, such as the p-adic field $\mathbb{C}_p$. In the complex world we get every Riemann surface as a quotient of an open subset Ω of $\mathbb{P}^1(\mathbb{C})$ by a discrete subgroup G of $\mathrm{PSL}_2(\mathbb{C})$. In the p-adic world the analogues of Riemann surfaces, which admit a similar uniformization Ω/G, are called *Mumford curves*. But contrary to the complex world not every nonsingular projective curve over $\mathbb{C}_p$ is a Mumford curve. Mumford curves have been thoroughly studied; two textbooks giving a comprehensive introduction are [GP] and [FP].

As Mumford curves are the p-adic analogues of Riemann surfaces we define p-adic origamis to be coverings of Mumford curves with only one branch point, where the bottom curve has genus one. In the fourth chapter we will classify all normal non-trivial p-adic origamis. This is done using the description of the bottom curve as an orbifold Ω/G, where $\Omega \subset \mathbb{P}^1(\mathbb{C}_p)$ and G is a group acting discontinuously on Ω. These groups and the corresponding orbifolds can be studied by looking at the action of G on the Bruhat-Tits-Tree of a suitable subfield of $\mathbb{C}_p$ and the resulting quotient graph of groups. This has been done by Herrlich [Her], and more recently by Kato [Kat2] and Bradley [Bra2].

In Section 4.3 we will see that all normal p-adic origamis with a given Galois group H are of the type $\Omega/\Gamma \to \Omega/G$ with the following possible choices for the groups Γ and G: The quotient graph of G can be contracted to

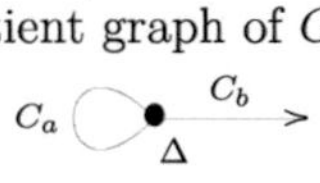

for a p-adic triangle group Δ (where the single vertex represents a subtree with fundamental group Δ), which means that G is isomorphic to the fundamental group of this graph, i.e.

$$G \cong \langle \Delta, \gamma;\, \gamma\alpha_1 = \alpha_2\gamma \rangle \text{ with } \alpha_i \in \Delta \text{ of order } a.$$

Γ is the kernel of a morphism $\varphi : G \to H$ which is injective when restricted to the vertex groups of the quotient graph of G. The ramification index of the p-adic origami is then b. We have a similar result (Theorem 4.18) for the automorphism group of the p-adic origami.

Given a p-adic origami which is defined over $\overline{\mathbb{Q}}$ we can change the ground field to $\mathbb{C}$ and know that there our origami can be described as a surface glued from squares. Actually doing this is usually hard, because we would have to work out equations for the Mumford curves and for the complex curves corresponding to the Riemann surfaces. Nevertheless we can often find out which complex origami-curve belongs to our p-adic origami, as mostly the curve is already uniquely defined by fixing the Galois group. In the last chapter we prove that this is true for the Galois groups $D_n \times \mathbb{Z}/m\mathbb{Z}$ and $A_4 \times \mathbb{Z}/m\mathbb{Z}$ (with $n, m \in \mathbb{N}$ and n odd).
We will also discuss some cases where this does not work, i.e. where there are several origami-curves of origamis with the same Galois group. In Appendix A.3 we see that for groups of order less than or equal to 250 this happens only for 30 groups. To construct examples we can take an origami-curve which is not fixed by an element σ of the absolute Galois group $\mathrm{Gal}(\overline{\mathbb{Q}}/\mathbb{Q})$. As the action of $\mathrm{Gal}(\overline{\mathbb{Q}}/\mathbb{Q})$ on origami-curves is faithful by [Möl], Th. 5.4 we can find such a curve for any given σ. In this case of course both the curve and its image contain origamis with the same Galois group, and we suspect that all other known invariants of origami-curves are equal as well.

Acknowledgments. Several people have supported me during my work in Karlsruhe. In particular I want to thank:
Professor Frank Herrlich for encouraging my work on this thesis and supporting me throughout the whole time. Especially for giving me the freedom to work in my own pace while at the same time hinting (more or less subtle) that the thesis might be finished soon. Also for the fast proofreading of the manuscript and his helpful comments. *Stefan Kühnlein* also for proofreading and for kindly agreeing to be Korreferent of this thesis. *Patrick Erik Bradley* for introducing me to Kato's work on quotient graphs by handing me an early draft of [BKV] and for proofreading the p-adic part. *Gabriela Schmithüsen* for a copy of her excellent thesis [Sch1] (which by now has become totally disconnected) and for letting me redesign her source codes of the algorithms therein. Also for proofreading the complex part and founding the research project *"With origamis to Teichmüller curves in moduli space"*, whose other members *André Kappes, Florian Nisbach* and *Myriam Finster* also helped with many fruitful discussions about origamis and testing my algorithms. *Oliver Bauer, Sven Heumann* and *Susanne Schmitt*, who shared (and partly still share) an office with me, for the very friendly (often hilarious) working atmosphere. *Hendrik Kasten, Ute Wolf, Fabian Januszewski, Sebastian Rit-*

terbusch, Marc Mitschele and *Mathias Krause* for their support and many discussions about various mathematical topics.

Furthermore I thank the *Landesgraduiertenförderung* of Baden-Württemberg for providing a grant, and the workgroups *Number Theory and Algebraic Geometry*, *Inverse Problems* and *Differential Geometry*, as well as the *Chair of Scientific Computing* of the University of Karlsruhe for giving me the opportunity to work and teach there.

Karlsruhe, June 2009 Karsten Kremer

Contents

Chapter 1

Origamis

In this first chapter we want to introduce origamis from a variety of viewpoints. Origamis are compact Riemann surfaces which are obtained by glueing several unit squares. Pierre Lochak coined the term *origami* in [Loc], while Anton Zorich uses the term *square-tiled surfaces* [Zor]. But there are several other ways to define origamis: Origamis can also be defined as coverings of an elliptic curve with a single branch point, which is a concept one can generalize to algebraic curves over other fields than $\mathbb{C}$. In [Sch1] Gabriela Schmithüsen often identifies origamis with conjugacy classes of finite index subgroups of F_2. If one wants to use a computer for the calculation of some properties of origamis, it is more practical to define origamis as homomorphisms $F_2 \to S_d$, or (in the case of normal origamis with Galois group G) as epimorphisms $F_2 \to G$. But if one thinks of origamis as special kinds of translation surfaces it may be more natural to think of an origami as a Riemann surface with a holomorphic 1-form which has integer values when integrated along elements of the fundamental group.

We will now explain all those definitions, prove that they are equivalent, and investigate how the natural action of $\mathrm{SL}_2(\mathbb{Z})$ on the set of all origamis can be described in each case.

1.1 Definitions

Definition 1.1. An *origami* of degree $d \in \mathbb{N}$ is a closed surface X which is obtained from d euclidean unit squares by glueing (via translations) each right edge to a left one and each upper edge to a lower one.

By labeling the squares of an origami with the numbers $1, \ldots, d$ we can describe an origami by two permutations σ_x and σ_y in S_d, where σ_x describes the horizontal glueing and σ_y describes the vertical glueing.

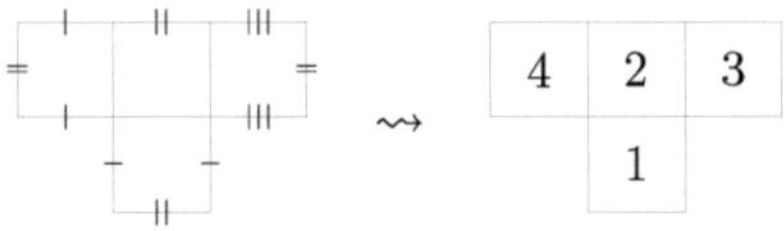

Example: This origami is described by $\sigma_x = (2\,3\,4)$ and $\sigma_y = (1\,2)$.

Each origami X defines a covering of the torus $E := \mathbb{R}^2/\mathbb{Z}^2$ by mapping each square to the unit square. This covering $p : X \to E$ is ramified only at the vertices of the squares. Removing these ramification points leads to an unramified restriction $p : X^* := X \setminus p^{-1}(0) \to E^* := E \setminus \{0\}$ of degree d.

Conversely, given a connected surface X and a covering $p : X \to E$ of degree d which may be ramified only over $0 \in E$, we can lift the unit square defining E to X. This yields a decomposition of X into d copies of the unit square glued as described above.

The monodromy of such a covering is by definition the action[1] of the fundamental group $\pi_1(E^*, P)$ on the fiber $p^{-1}(P)$ over any basepoint $P \in E^*$, and without loss of generality we can choose both coordinates of P to be non-zero in $\mathbb{R}/\mathbb{Z}$. The fundamental group $\pi_1(E^*, P)$ is isomorphic to the free group generated by x and y, where x is the closed path starting at P in horizontal direction, and y is the closed path starting at P in vertical direction. Let P_i be the preimage of P in the square with the number i. Then we see that the monodromy of the origami is given by the homomorphism $f : F_2 \to \mathrm{Sym}(\{P_1, \ldots, P_d\}) \cong S_d$ which maps x to σ_x and y to σ_y.

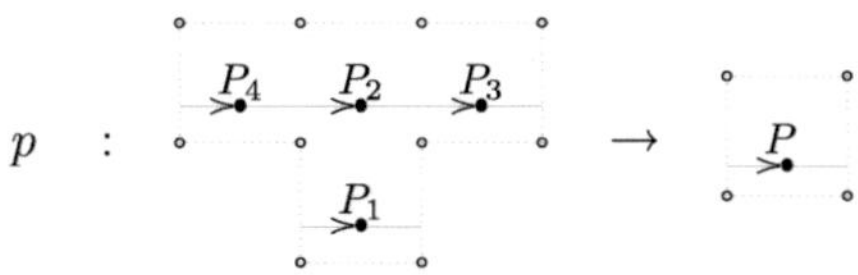

On the other hand a homomorphism $f : F_2 \to S_d$ describes an origami iff its image is a transitive subgroup of S_d (otherwise the surface obtained by glueing the squares $1, \ldots, d$ according to the permutations $f(x)$ and $f(y)$ would not be connected). We denote by $\mathrm{Hom}^t(F_2, S_d)$ the set of such homomorphisms.

[1] Note that if we want to consider elements $\alpha, \beta \in \pi_1(E^*, P)$ as permutations of the fiber $p^{-1}(P)$, we need $\alpha\beta$ to be the path *first* along β and *afterwards* along α. This may not be an intuitive way to define multiplication in $\pi_1(E^*, P)$, but otherwise the group would not act on the fiber from the left.

The unramified covering $p : X^* \to E^*$ also induces an inclusion ι of the fundamental groups $\pi_1(X^*, P_1) \hookrightarrow \pi_1(E^*, P) \cong F_2$. Let H denote the image of this inclusion in F_2. The left cosets of H correspond to the d different squares, which make up X^*, thus the index of H in F_2 is d. The group H can also be obtained from the monodromy f: as H is just the group of all words in F_2 which describe a path from P_1 back to P_1 we can write $H = f^{-1}(\mathrm{Stab}_{S_d}(1))$.

Conversely, given a subgroup $H \subseteq F_2$ of finite index d, we can define an origami in the following way: We label d squares with the left cosets of H. For every square labelled aH we glue its right edge to the left edge of xaH, and its upper edge to the lower edge of yaH.

Altogether we can deduce

Proposition 1.2. *An origami of degree d can be defined equivalently as*

i) a finite covering $p : X \to E$ of the torus E by a connected surface X, ramified only over $0 \in E$; up to a homeomorphism $X' \to X$ over E,

ii) an element of $\mathrm{Inn}(S_d)\backslash \mathrm{Hom}^t(F_2, S_d)$,

iii) the conjugacy class of a subgroup H of F_2 of index d.

Proof. We already described above how coverings $p : X \to E$ of degree d, homomorphisms $f : F_2 \to S_d$ and subgroups $H \subseteq F_2$ of index d are connected to our definition of an origami. Now we just have to factor out the choices we made:

i) For another connected surface X' let $h : X' \to X$ be a homeomorphism and consider the finite covering $p' := p \circ h$. As h is unramified the monodromies of p and p' coincide if we set $P_i' := h^{-1}(P_i)$.

ii) When we label the squares with the numbers $1, \ldots, d$ any other numbering leads to the same origami. This means, that two monodromy maps f and f' describe the same origami, iff there is a renumbering $\tau \in S_d$ with $f(x) = \tau f'(x)\tau^{-1}$ and $f(y) = \tau f'(y)\tau^{-1}$. This means $f = \kappa_\tau \circ f'$, where $\kappa_\tau \in \mathrm{Inn}(S_d)$ is the conjugation with τ.

iii) For the definition of the subgroup H we only choose the basepoint P_1 of the fundamental group $\pi_1(X^*)$ out of the d elements in the fiber $p^{-1}(P)$. If we choose another basepoint P_i instead, then we have to conjugate each element in H with the image of the path from P_1 to P_i in $\pi_1(E^*)$.

□

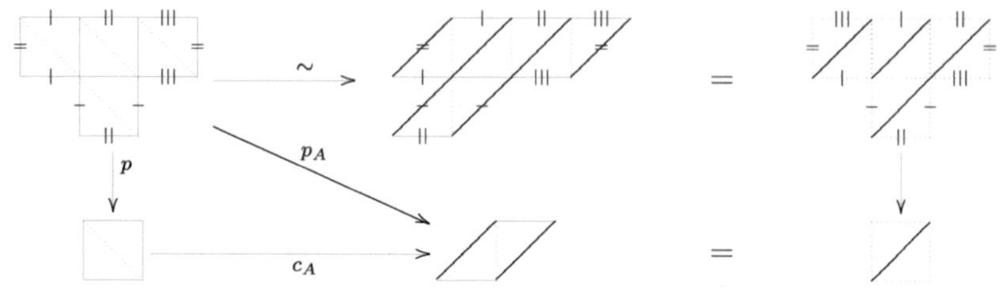

Figure 1.1: The image of an origami under the action of $\mathrm{SL}_2(\mathbb{Z})$

We remark (to part ii) of the previous proposition) that $\mathrm{Inn}(S_d) = \mathrm{Aut}(S_d)$ for $d \neq 6$ (see [JR] Theorem 7.4), but there exist automorphisms of S_6 which are not inner (see [Rot] Theorem 3).

1.2 Deformation of origamis

We identify the torus $E = \mathbb{R}^2/\mathbb{Z}^2$ with $\mathbb{C}/\mathbb{Z}[i]$, thus an origami $p : X \to E$ becomes a Riemann surface using the coordinate charts induced by p. In fact, we get a lot of Riemann surfaces: for every $A \in \mathrm{SL}_2(\mathbb{R})$ we can define the lattice $\Lambda_A = A \cdot \mathbb{Z}^2$ and the homeomorphism $c_A : \mathbb{R}^2/\mathbb{Z}^2 \to \mathbb{R}^2/\Lambda_A =: E_A, x \mapsto A \cdot x$. The identification of $\mathbb{R}^2$ with $\mathbb{C}$ then leads to new coordinate charts induced by $p_A := c_A \circ p$. We get again a complex structure on the surface X which we denote by X_A.

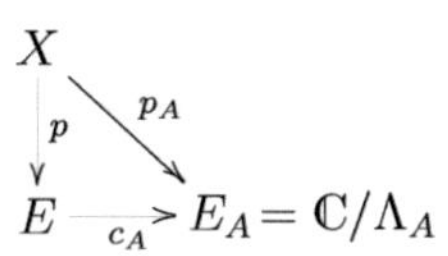

If the torus E_A is isomorphic to our torus $E = E_I$ as a Riemann surface, then $p_A : X \to E_A \cong E$ defines another origami. By

$$E_A = E \Leftrightarrow \Lambda_A = \Lambda \Leftrightarrow A \in \mathrm{SL}_2(\mathbb{Z})$$

we see that we get an action of $\mathrm{SL}_2(\mathbb{Z})$ on the set of all origamis.

By [LS], I.4.5 we have $\mathrm{GL}_2(\mathbb{Z}) \cong \mathrm{Out}(F_2) := \mathrm{Aut}(F_2)/\mathrm{Inn}(F_2)$. Let $\mathrm{Out}^+(F_2)$ be the subgroup corresponding to $\mathrm{SL}_2(\mathbb{Z})$ and $\mathrm{Aut}^+(F_2)$ be its preimage in $\mathrm{Aut}(F_2)$. This enables us to formulate the action of $\mathrm{SL}_2(\mathbb{Z})$ on origamis also for homomorphisms $F_2 \to S_d$ and for finite-index subgroups H of F_2:

Proposition 1.3. *$A \in \mathrm{SL}_2(\mathbb{Z}) \cong \mathrm{Out}(F_2) = \mathrm{Aut}(F_2)/\mathrm{Inn}(F_2)$ acts on origamis in the following way (where $\varphi \in \mathrm{Aut}(F_2)$ is a preimage of A):*

i) $p \mapsto p_A = c_A \circ p$ *for a finite covering* $p : X \to E$,

ii) $[f] \mapsto [f \circ \varphi^{-1}]$ *for* $[f] \in \mathrm{Inn}(S_d)\backslash \mathrm{Hom}^t(F_2, S_d)$,

iii) $[H] \mapsto [\varphi(H)]$ *for the conjugacy class of a subgroup* H *of* F_2.

Proof. We have to investigate how the action described above, leading to statement i), induces the actions stated in ii) and iii). Recall from Prop. 1.2 that both alternative definitions of origamis which we use for ii) and iii) were obtained via the fundamental group $\pi_1(E^*, P)$. By the theorem of Dehn-Nielsen the automorphism φ of $\pi_1(E^*, P)$ comes from a unique homotopy class of homeomorphisms c of E^*. The isomorphism

$$\mathrm{Homeo}(E^*)/\mathrm{Homeo}^0(E^*) \cong \mathrm{SL}_2(\mathbb{Z}) \cong \mathrm{Out}^+(F_2)$$

is in fact the map $c \mapsto [\varphi]$, so we have $c = [c_A]$. The induced maps on the fundamental groups and the monodromy maps are shown in the following diagram:

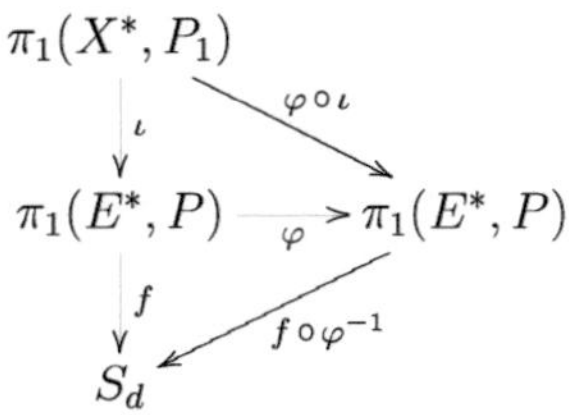

Thus the monodromy f transforms to $f \circ \varphi^{-1}$, and the subgroup $H = \mathrm{Im}(\iota)$ to $\mathrm{Im}(\varphi \circ \iota) = \varphi(H)$. □

1.3 Normal origamis

Proposition 1.4. *The following statements are equivalent:*

i) $p : X \to E$ *is a normal covering,*

ii) H *is a normal subgroup of* F_2,

iii) $H = \ker(f)$.

Proof. The covering p is normal if and only if $\pi_1(X^*, P_1) \lhd \pi_1(E^*, P)$. This shows the equivalence i) $\Leftrightarrow$ ii). The kernel of f is always a normal subgroup of F_2, hence iii) $\Rightarrow$ ii).

It remains to show ii) $\Rightarrow$ iii): Starting from a finite index subgroup H we have constructed the corresponding surface X in Section 1.1 by labeling d squares with the left cosets of H. If $H \lhd F_2$ the monodromy is given by the canonical morphism $f : F_2 \to F_2/H$. □

Definition 1.5. An origami which satisfies one of the conditions of Proposition 1.4 is called *normal*. The group $F_2/H \cong \mathrm{Im}(f)$ is called its *Galois group*.

Proposition 1.6. *A normal origami with Galois group G can equivalently be defined as an element of* $\mathrm{Aut}(G)\backslash\,\mathrm{Epi}(F_2, G)$. *The action of* $\mathrm{Out}(F_2)$ *on such origamis is given by* $[\varphi] \cdot [f] = [f \circ \varphi^{-1}]$.

Proof. For a given epimorphism $f : F_2 \to G$ we use $H := \ker(f)$ as a finite index subgroup of F_2 to define an origami. This kernel is invariant under automorphisms of G.

On the other hand let H be a finite index normal subgroup of F_2. We have seen in the proof of Prop. 1.4 that the monodromy map of the corresponding origami can be written as a surjective homomorphism $f : F_2 \to F_2/H \cong G$ (which is of course unique only up to the chosen isomorphism $F_2/H \cong G$). If we start with another representative of the conjugacy class of H, this changes our map only by an inner automorphism of G. Note that indeed $H = \ker(f)$, therefore this construction is inverse to the one described above.

If we use the interpretation of $f \in \mathrm{Epi}(F_2, G)$ as the monodromy map of our origami, then Prop. 1.3 ii) describes the action of $\mathrm{Out}(F_2)$. □

1.4 Abelian Differentials

Origamis are special cases of translation surfaces and as such closely related to Abelian differentials. We give here a short introduction along the lines of [Zor], Sections 2 and 4:

Definition 1.7. A *translation surface* is a Riemann surface X together with a finite set $S := \{P_1, \ldots, P_n\} \subset X$ of *singularities* and an atlas such that

i) On $X^* := X \setminus S$ all coordinate-change maps are translations.

ii) For each $P \in S$ there is a chart $f : U \to \mathbb{C}$ with $f(P) = 0$ such that every coordinate-change map $f(U) \to \mathbb{C}$ is of the form $z \mapsto z^k$ for a $k \in \mathbb{N}_{>1}$, called the *multiplicity* of the singularity P.

On a translation surface X two charts z and z' on X^* differ only by a constant. Therefore the locally defined 1-form dz equals dz'. For the additional charts with coordinate-change maps $z \mapsto z^k$ as above we can use $dz' = kz^{k-1}dz$ (which has a zero $z = f(P) = 0$ of order $k-1$) to extend our 1-form holomorphically to neighborhoods of S. Therefore by setting $dz = 0$ on S we can extend dz to a globally defined holomorphic 1-form $\omega \in H^0(X, \Omega_X)$. Such a 1-form is also called *Abelian differential.*

This works in the other direction as well: Given an Abelian differential on a surface X with zeroes $P_1, \ldots, P_n$, we can reconstruct the translation structure on $X^* := X \setminus \{P_1, \ldots, P_n\}$ using locally the charts $U \to \mathbb{C}; x \mapsto \int_{x_0}^{x} \omega$ for any simply connected $U \subset X^*$ with arbitrary $x_0 \in U$ (different charts differ only by a constant).

Proposition 1.8. *An Abelian differential ω on a Riemann surface X defines an origami if and only if $\int_\gamma \omega \in \mathbb{Z} + i\mathbb{Z}$ for every $\gamma \in \pi_1(X^*)$ and for every path γ connecting two zeroes of ω.*

In this case the zeroes of ω of order $k-1$ are the ramification points of the origami $X \to E$ with ramification index k.

Proof. Let $p : X \to E$ be an origami. As a Riemann surface glued from unit squares by translations X is a translation surface with singularities at the vertices of the squares, i.e. the ramification points of p. At a ramification point of order k we have coordinate-change maps $z \mapsto z^k$, therefore ω has a zero of order $k-1$ at this point.

We can calculate the given integrals on E:

$$\int_\gamma \omega = \int_{p \circ \gamma} dz$$

where $p \circ \gamma$ is a closed path on $E = \mathbb{C}/(\mathbb{Z} + i\mathbb{Z})$. Let $\overline{\gamma}$ be a lift of $p \circ \gamma$ in $\mathbb{C}$. Its endpoints are equal modulo $\mathbb{Z} + i\mathbb{Z}$. Therefore $\int_{\overline{\gamma}} dz \in \mathbb{Z} + i\mathbb{Z}$.

On the other hand, given an Abelian differential with the stated properties for any $z_0 \in X^*$ the map

$$p : X \to \mathbb{C}/(\mathbb{Z} + i\mathbb{Z});\ z \mapsto \int_{z_0}^{z} \omega$$

is well-defined and ramified only at the zeroes of ω, which are preimages of $0 \in E$. □

In [Zor] the statement of Proposition 1.8 occurs in the context of period coordinates of Abelian differentials, which we will introduce as in [EO], §1.2: For an origami $p : X \to E$ we will usually denote the set of (possible) ramification points $p^{-1}(0)$ by $R = \{R_1, \ldots, R_n\}$ (instead of $S = \{P_1, \ldots, P_n\}$), and the ramification index at the point R_i by e_i (instead of k). There may be $R_i \in p^{-1}(0)$ where p is not ramified, hence $e_i = 1$ is also possible[2]. Setting $r_i = e_i - 1$ for each ramification point we note that the divisor K of ω equals $\sum r_i P_i$ with $\sum r_i = 2g - 2$ (using Riemann-Hurwitz).

Let $\mathscr{H}$ denote the moduli space of Abelian differentials with zeroes of order $e_1, \ldots, e_n$ at points $P_1, \ldots, P_n$ on a surface X of genus $g = 1 + \frac{1}{2}\sum(e_i - 1)$. Consider the relative homology group $H_1(X, \{P_1, \ldots, P_n\}, \mathbb{Z})$ and choose a standard basis $\{\gamma_i\}$ (i.e. a symplectic basis $\gamma_1, \ldots, \gamma_{2g}$ of $\pi_1(X^*)$ and paths γ_{2g+i-1} from P_1 to P_i for $i = 2, \ldots, n$). Now the map

$$\Phi : \mathscr{H} \to \mathbb{C}^n; \quad \omega \mapsto \left(\int\limits_{\gamma_1} \omega, \ldots, \int\limits_{\gamma_{2g+n-1}} \omega \right)$$

is called the *period map*. This map defines a local coordinate system on $\mathscr{H}$; the coordinates of $\omega \in \mathscr{H}$ are called its *period coordinates*. Proposition 1.8 states that all period coordinates of ω are integer if and only if the translation surface defined by ω is an origami. Therefore we can think of origamis as the integer points in $\mathscr{H}$.

[2]Note that this means that there are always several different origamis corresponding to a given Abelian differential ω, because subdividing the squares of an origami into smaller squares means adding points with $e_i = 1$ without changing ω.

Chapter 2

Origami-curves

The moduli space $\mathcal{M}_{g,n}$ is the set of isomorphism classes of Riemann surfaces of genus g with n punctures, endowed with the structure of an algebraic variety. An origami defines a curve in this moduli space via the deformation mentioned in section 1.2. In this chapter we will explain this construction, and then investigate under which conditions two origamis define the same curve in moduli space. We omit some details which can be found for instance in [HS2], Ch. 1 and 2. For those unfamiliar with the concept of Teichmüller spaces and moduli spaces we recommend [Nag] as an introduction.

2.1 From Teichmüller space to moduli space

We want to study the *moduli space*

$$\mathcal{M}_{g,n} := \left\{\text{compact Riemann surfaces } X \text{ of genus } g \text{ with } n \text{ punctures}^1\right\} /_{\sim}$$

with $X_1 \sim X_2$ if there is a biholomorphic map $h : X_1 \to X_2$.
This space has a structure of a complex algebraic variety of dimension $3g - 3 + n$ (we will consider only the case where this number is positive). But in general this variety has singularities wherever the group of automorphisms of the corresponding Riemann surface is non-trivial. This problem can be resolved by looking instead at the set of Riemann surfaces endowed with a so-called *marking* (which changes if an automorphism is applied, thus resulting in a different object in the classifying space). As marking we use $f \in \text{Diffeo}^+(X_{\text{ref}}, X)$, an orientation preserving diffeomorphism $X_{\text{ref}} \to X$

[1] A puncture of X is a marked point. A biholomorphic map $h : X_1 \to X_2$ has to map punctures of X_1 to punctures of X_2.

from a fixed Riemann surface X_{ref} of genus g with n punctures to our Riemann surface X, and thus arrive at the *Teichmüller space*

$$\mathcal{T}_{g,n} := \left\{(X, f) : X \text{ as above}, f \in \text{Diffeo}^+(X_{\text{ref}}, X)\right\} /_{\sim}$$

where $(X_1, f_1) \sim (X_2, f_2)$ if the map $f_2 \circ f_1^{-1} : X_1 \to X_2$ is homotopic to a biholomorphic map $h : X_1 \to X_2$.

We can map $\mathcal{T}_{g,n}$ to $\mathcal{M}_{g,n}$ by forgetting the marking f. A marking f of a Riemann surface X can be transformed into any other marking of X by concatenating f with an orientation preserving diffeomorphism of X_{ref}. If the diffeomorphism is homotopic to the identity then of course the equivalence class of (X, f) in $\mathcal{T}_{g,n}$ will not change. We denote the set of those diffeomorphisms by $\text{Diffeo}^0(X_{\text{ref}})$. This leads to the definition of the *mapping class group*

$$\Gamma_{g,n} := \text{Diffeo}^+(X_{\text{ref}}) / \text{Diffeo}^0(X_{\text{ref}})$$

$\Gamma_{g,n}$ acts properly discontinuously on $\mathcal{T}_{g,n}$ and the orbit space $\mathcal{T}_{g,n}/\Gamma_{g,n}$ is $\mathcal{M}_{g,n}$. One can define a metric (the so-called *Teichmüller-metric*, see [Nag], §2.1.7) and a complex structure on Teichmüller space, which turns $\mathcal{T}_{g,n}$ into a complex manifold of dimension $3g-3+n$ (by the Theorem in [Nag], §2.5.5).

Now let O be an origami defined by $p : X \to E$. Let g be the genus of X and $n := |p^{-1}(0)|$ the number of its punctures. For $A \in \text{SL}_2(\mathbb{R})$ we use the identity-map $\text{id} : X_I \to X_A$ to define a marking of X_A. With this marking X_A defines an element $[X_A]$ in the Teichmüller space $\mathcal{T}_{g,n}$. Therefore we have a map

$$\iota : \text{SL}_2(\mathbb{R}) \to \mathcal{T}_{g,n}, \ A \mapsto [X_A].$$

Let Δ_O be the image of this map, and $c(O)$ the image of Δ_O in $\mathcal{M}_{g,n}$.

Note that for $A \in \text{SO}_2(\mathbb{R})$ we have $(X, \text{id}) \sim (X_A, \text{id})$ because the map $X \to X_A$ is locally defined by $c_A : E \to E_A$, which is biholomorphic. Therefore ι is constant on each $\text{SO}_2(\mathbb{R})$ orbit and we get a map

$$\iota : \mathbb{H} \cong \text{SO}_2(\mathbb{R}) \backslash \text{SL}_2(\mathbb{R}) \to \mathcal{T}_{g,n}$$

and in fact we can choose the isomorphism $\mathbb{H} \cong \text{SO}_2(\mathbb{R}) \backslash \text{SL}_2(\mathbb{R})$ in such a way that $\iota : \mathbb{H} \to \mathcal{T}_{g,n}$ is a holomorphic isometric embedding (see [HS2], Definition 2.7). This fits into a more general context: In general the image of such a map is called a *Teichmüller disk*. If its image in $\mathcal{M}_{g,n}$ is an algebraic curve, then this curve is called a *Teichmüller curve*. For origamis this is always the case:

Proposition 2.1. *For an origami O the set $c(O)$ is an algebraic curve.*

This is proven in [Loc], Prop. 3.2 ii). We call such a curve $c(O)$ an *origami-curve*. Furthermore Lochak proves there that $c(O)$ is defined over a number field. We will sketch a proof for this in Remark 2.5.

2.2 Counting origami-curves

Given two origamis O and O' we would like to check whether both origamis define the same origami-curve or not. It is helpful to know

Proposition 2.2. $c(O) = c(O')$ *if and only if* O *and* O' *are in the same* $\mathrm{SL}_2(\mathbb{Z})$ *orbit,*

which is proven in [HS1], Prop. 5 b) for the definition of the $\mathrm{SL}_2(\mathbb{Z})$ action of Prop. 1.3 iii).

We would now like to know how many origami-curves exist for a given degree d of the origamis. By Proposition 1.2 ii) an origami can be represented by a transitive homomorphism $f : F_2 \to S_d$, which is defined by $\sigma_x := f(x)$ and $\sigma_y := f(y)$ in S_d. Proposition 1.3 ii) tells us that the $\mathrm{SL}_2(\mathbb{Z})$ action in this case corresponds to the action of $\mathrm{Out}(F_2)$ by concatenation. Thus we want to calculate the cardinality of the set

$$\mathrm{Inn}(S_d)\backslash \mathrm{Hom}^t(F_2, S_d)/ \mathrm{Out}(F_2)$$

To do this we first omit the condition 'transitive' and calculate coset representatives. Afterwards we count only those representatives which define a transitive homomorphism.

For the calculation of the coset representatives we choose a set $\mathrm{Gen}(\mathrm{Out}(F_2))$ of lifts of generators of $\mathrm{Out}(F_2)$ in $\mathrm{Aut}(F_2)$ and a set $\mathrm{Gen}(\mathrm{Inn}(S_d))$ of generators of $\mathrm{Inn}(S_d)$. Then we define the graph $\mathcal{G} = (V, E)$ with vertex set $V = \mathrm{Hom}(F_2, S_d)$ (which can be represented by $S_d \times S_d$) and edge set

$$E = \{\{f, f \circ \varphi\} : \varphi \in \mathrm{Gen}(\mathrm{Out}(F_2))\} \cup \{\{f, \psi \circ f\} : \psi \in \mathrm{Gen}(\mathrm{Inn}(S_d))\}$$

Two vertices in this graph are connected if and only if they correspond to origamis in the same $\mathrm{SL}_2(\mathbb{Z})$ orbit. Thus we get our coset representatives by picking one vertex out of every connected component of the graph $\mathcal{G}$.

As generating set of $\mathrm{Out}(F_2)$ we use $\{[\varphi_S], [\varphi_T]\}$ with

$$\varphi_S(x) = y,\ \varphi_S(y) = x^{-1},\ \varphi_T(x) = x,\ \varphi_T(y) = xy$$

which corresponds to the generating set $\{S, T\}$ of $\mathrm{SL}_2(\mathbb{Z})$ given by

$$S = \begin{pmatrix} 0 & -1 \\ 1 & 0 \end{pmatrix} \text{ and } T = \begin{pmatrix} 1 & 1 \\ 0 & 1 \end{pmatrix}.$$

As generating set of $\mathrm{Inn}(S_d)$ we use $\left\{\kappa_{(1\,i)} : i \in \{1, \ldots, d\}\right\}$ where $\kappa_{(1\,i)}$ is the conjugation with the transposition $(1\ i) \in S_d$.

An implementation of this algorithm can be found in appendix A.1. We can use it to calculate the number of origami-curves in the moduli spaces $\mathcal{M}_{g,n}$.

degree d	genus g	n punctures	number of curves
1	1	1	1
2	1	2	1
3	1	3	1
	2	1	1
4	1	4	2
	2	2	3
5	1	5	1
	2	3	3
	3	1	4
6	1	6	1
	2	4	8
	3	2	19
7	1	7	1
	2	5	4
	3	3	22
	4	1	14

A more detailed examination of properties of these origami-curves follows in Appendix A.2. Note that we have to compute the connected components of a graph with $(d!)^2$ vertices, for $d = 7$ these are already more than 25 million. Thus the memory usage grows exponentially in d, therefore for large d it won't be possible to use this algorithm.

If we consider only normal origamis, then we can modify the algorithm such that it works for large origamis as well: If we fix a Galois group G, then each origami is represented by an epimorphism $F_2 \to G$. The set of origami-curves containing normal origamis with Galois group G therefore corresponds to

$$\mathrm{Aut}(G) \backslash \mathrm{Epi}(F_2, G) / \mathrm{Out}(F_2)$$

We can thus replace S_d in our algorithm by G, the transitivity of the image of f by its surjectivity and $\mathrm{Gen}(\mathrm{Inn}(S_d))$ by a generating set of $\mathrm{Aut}(G)$. An implementation can also be found in appendix A.1. Note that for the graph we use now the vertex-set $G \times G$, and the cardinality of G is the degree d of our origami. Therefore the memory usage grows only quadratically in d.

Example 2.3. We use this algorithm to calculate for example the number of origami-curves containing normal origamis with Galois group $\mathrm{PSL}(2,9)$:

degree d	genus g	n punctures	number of curves
660	166	330	1
	221	220	1
	265	132	4
	276	110	2
	301	60	1

2.3 The Veech group

Let X be a translation surface. An automorphism f of X is called *affine*, if it is (at least on the punctured surface X^*) locally defined by maps $z \mapsto A \cdot z + b$ with $b \in \mathbb{C}$ and $A \in \mathrm{GL}_2(\mathbb{R})$ acting on $z \in \mathbb{C}$ by Moebius transformation. Let $\mathrm{Aff}^+(X)$ denote the group of orientation preserving affine diffeomorphisms of X. As X is a translation surface the coordinate change maps of X^* are translations. Therefore A is independent of the chosen coordinate charts and thus also independent of the chosen neighborhood. We call this matrix A the *derivative* $\mathrm{der}(f)$ of the affine diffeomorphism f. Note that on a closed surface f has to be area preserving, thus (together with orientation preserving) we get $\mathrm{der}(f) \in \mathrm{SL}_2(\mathbb{R})$. The image $\mathrm{der}(\mathrm{Aff}^+(X)) \subset \mathrm{SL}_2(\mathbb{R})$ of the derivative map is called the *Veech group* $\Gamma(X)$ of X (introduced by Veech in [Vee]).

For origamis we also have another more accessible characterization of the Veech group:

Proposition 2.4. *The Veech group of an origami is the stabilizer in* $\mathrm{SL}_2(\mathbb{Z})$ *under the action defined in Proposition 1.3.*

This was shown in [Sch2], Prop. 1 for the definition stated in Prop. 1.2 iii). It implies that for an origami O there is a one-to-one correspondence of the left cosets of the Veech group $\Gamma(O)$ in $\mathrm{SL}_2(\mathbb{Z})$ and the orbit of O under the $\mathrm{SL}_2(\mathbb{Z})$ action, and by Prop. 2.2 this corresponds to the set of origamis O' which describe the same curve $c(O') = c(O)$ in moduli space.

Note that the set of origamis of degree d is a finite set (as quotient of the finite set $\mathrm{Hom}^t(F_2, S_d)$). Therefore the Veech group (as the stabilizer under this action) is a subgroup of finite index in $\mathrm{SL}_2(\mathbb{Z})$. Gutkin and Judge have proven in [GJ], Theorem 5.5 that having a Veech group commensurable to $\mathrm{SL}_2(\mathbb{Z})$ is a property which characterizes origamis.

Remark 2.5. *Let O be an origami and $\overline{\Gamma}(O)$ the image of $\Gamma(O)$ in* $\mathrm{PSL}_2(\mathbb{Z})$. *Then $\overline{\Gamma}(O)\backslash\mathbb{H}$ is an algebraic curve, and the map*

$$\overline{\Gamma}(O)\backslash\mathbb{H} \to \mathrm{PSL}_2(\mathbb{Z})\backslash\mathbb{H} \cong \mathbb{P}^1(\mathbb{C})$$

is ramified at most over three points, therefore by Belyi's theorem the curve is defined over a number field (a fact we remarked already after Prop. 2.1). If we replace $\overline{\Gamma}(O)$ by its conjugate with the map $z \mapsto -\bar{z}$, then the resulting curve is actually the normalization of the origami-curve $c(O)$ in $\mathcal{M}_g$ (see [HS2], Cor. 2.21, or for a detailed proof [Mc1], Cor. 3.3)

The Veech group also leads to a simple algorithm for listing all origamis which describe the same curve: Given an origami O defined by $f \in \mathrm{Hom}^t(F_2, S_d)$ first calculate a set of left coset representatives R of the Veech group Γ of O. Now every element $g \in \mathrm{SL}_2(\mathbb{Z})$ can be written as $g = r\gamma$ with $r \in R$ and $\gamma \in \Gamma$. As γ stabilizes O the image gO equals rO. Therefore the set $\{rO : r \in R\}$ equals the orbit of O under the action of $\mathrm{SL}_2(\mathbb{Z})$ and thus contains all origamis describing the same origami-curve as O.

Constructing a set R of coset representatives is the most time-consuming part: We start with $R = \{\mathrm{id}\}$. Then for every element $x \in R$ we check for every generator g of $\mathrm{SL}_2(\mathbb{Z})$ (and for every inverse) whether the coset of gx is represented by an element in R (i.e. whether $gx\Gamma = y\Gamma$ for some $y \in R$, which is equivalent to $y^{-1}gx \in \Gamma$). If not then we have found a new coset and add it to the set R, then start all over again. As there are only finitely many cosets this algorithm terminates. Then a representative of every $x \in \mathrm{SL}_2(\mathbb{Z})$ is contained in R which we can prove by writing x as a product of the chosen generators and using induction over the word length.

This algorithm (based on the Reidemeister-Schreier method outlined in [LS], II.4) was already proposed by Schmithüsen in [Sch2] for the calculation of coset representatives (and generators) of the Veech group of origamis. Her algorithm calculates right coset representatives of the projective Veech group $\overline{\Gamma}$, which differs only marginally from the one presented here.

As generators of $\mathrm{SL}_2(\mathbb{Z})$ we use the elements S and T already mentioned in Section 2.2. Note that we get the set of coset representatives as words in the generators S and T, therefore we can easily write down a set of lifts $\{\varphi_1, \ldots, \varphi_m\}$ of those coset representatives in $\mathrm{Aut}(F_2)$ by replacing S by φ_S and T by φ_T (with φ_S and φ_T as defined on page 11). The homomorphisms $f \circ \varphi_i^{-1}$ then define all origamis with the same origami-curve $c(O)$.

For checking whether two origamis O and O' describe the same curve we just have to check for each φ_i whether $[f'] = [f \circ \varphi_i^{-1}]$ in $\mathrm{Inn}(S_d)\backslash \mathrm{Hom}^t(F_2, S_d)$, i.e. whether there is a $\sigma \in S_d$ such that $f \circ \varphi_i^{-1} = \kappa_\sigma \circ f'$ where κ_σ is the conjugation with σ. If there is no φ_i for which we can find such a σ, then the curves $c(O)$ and $c(O')$ cannot be equal.

Example 2.6. We illustrate this algorithm for the origami O defined by $\sigma_x = (1\,2)$ and $\sigma_y = (1\,3)$. We use the generators φ_S and φ_T of $\mathrm{Out}(F_2)$ and we are going to need their inverses:

$$\varphi_S^{-1}(x) = y^{-1},\ \varphi_S^{-1}(y) = x,\ \varphi_T^{-1}(x) = x,\ \varphi_T^{-1}(y) = x^{-1}y.$$

We then can calculate the images of our origami under the action of $\mathrm{SL}_2(\mathbb{Z})$:

$$\begin{array}{rlll}
f = & \begin{cases} x \mapsto (1\,2) \\ y \mapsto (1\,3) \end{cases} & & R := \{\mathrm{id}\} \\
f_S := f \circ \varphi_S^{-1} = & \begin{cases} x \mapsto f(y^{-1}) = (1\,3) \\ y \mapsto f(x) = (1\,2) \end{cases} & = \kappa_{(2\,3)} \circ f & S \in \Gamma \\
f_T := f \circ \varphi_T^{-1} = & \begin{cases} x \mapsto f(x) = (1\,2) \\ y \mapsto f(x^{-1}y) = (1\,3\,2) \end{cases} & & R := \{\mathrm{id}, T\} \\
f_{ST} := f_T \circ \varphi_S^{-1} = & \begin{cases} x \mapsto f_T(y^{-1}) = (1\,2\,3) \\ y \mapsto f_T(x) = (1\,2) \end{cases} & & R := \{\mathrm{id}, T, ST\} \\
f_{T^2} := f_T \circ \varphi_T^{-1} = & \begin{cases} x \mapsto f_T(x) = (1\,2) \\ y \mapsto f_T(x^{-1}y) = (1\,3) \end{cases} & = f & T^2 \in \Gamma \\
f_{S^2T} := f_{ST} \circ \varphi_S^{-1} = & \begin{cases} x \mapsto f_{ST}(y^{-1}) = (1\,2) \\ y \mapsto f_{ST}(x) = (1\,2\,3) \end{cases} & = \kappa_{(1\,2)} \circ f_T & T^{-1}S^2T \in \Gamma \\
f_{TST} := f_{ST} \circ \varphi_T^{-1} = & \begin{cases} x \mapsto f_{ST}(x) = (1\,2\,3) \\ y \mapsto f_{ST}(x^{-1}y) = (2\,3) \end{cases} & = \kappa_{(1\,3\,2)} \circ f_{ST} & (ST)^{-1}TST \in \Gamma
\end{array}$$

Thus the curve $c(O)$ contains exactly the three origamis defined by f, f_S and f_{ST}. This is illustrated in the following diagram:

Chapter 3

Invariants

In the previous chapter we have seen algorithms which decide whether two given origamis describe the same origami-curve in moduli space, i.e. whether the defining homomorphisms $F_2 \to S_d$ are equal in

$$\mathrm{Inn}(S_d) \backslash \mathrm{Hom}^t(F_2, S_d) / \mathrm{Out}(F_2).$$

Now we want to study data which are invariant for all origamis describing a common curve. Such invariants can be used to distinguish different curves. Of course the obvious invariant would be simply the full $\mathrm{SL}_2(\mathbb{Z})$ orbit of the origami, or equivalently listing the full coset in $\mathrm{Hom}^t(F_2, S_d)$. But this is based on the description of an origami glued from (complex) unit squares (or equivalently the monodromy homomorphism). In the following chapters we also want to consider origamis defined over other fields than $\mathbb{C}$ where we are not able to rely on this complex analytic characterization of origamis. Instead we will focus on invariants which represent properties of the covering map $X \to E$ and properties of the origami-curve itself.

3.1 Ramification indices

In this chapter O will always be an origami of degree d and genus g defined by the covering $p : X \to E$. Let $R := p^{-1}(0) = \{R_1, \ldots, R_n\}$ be the set of (possible) ramification points and let e_i denote the ramification index of p at R_i for $i = 1, \ldots, n$. Let $P \in E^* = E \setminus \{0\}$ be an arbitrary point and $\{P_1, \ldots, P_d\} = p^{-1}(P)$ be the fiber over P. Other notations from the first two chapters will also always refer to this fixed origami O if nothing else is indicated.

Proposition 3.1. *The degree d, the number of punctures n, the genus g and the ramification indices (e_i) of an origami are invariant under the action of* $\mathrm{SL}_2(\mathbb{Z})$.

Proof. For $A \in \mathrm{SL}_2(\mathbb{Z})$ the map c_A (defined in Section 1.2) is an unramified covering of degree 1, which fixes the point $0 \in E$. Therefore composing p with c_A neither changes the degree of the covering, nor the preimages of 0, nor the ramification index at any point of X. Of course the genus of X also remains unchanged. □

The tuple (e_i) is (up to changing the order of its entries) an invariant of the origami-curve. The numbers d, n and g are already coded in this invariant: For an origami with n ramification indices (e_i) we have $d = \sum e_i$ and $g = 1 + \frac{1}{2}(d - n)$, which follows from Riemann-Hurwitz or by calculating the Euler-characteristic.

Proposition 3.2. *For an origami with monodromy $f \in \mathrm{Hom}^t(F_2, S_d)$ set $\kappa := f(xyx^{-1}y^{-1})$. Then the ramification indices (e_i) equal the cycle-lengths of κ.*

Proof. Let $R \in p^{-1}(0)$ be a ramification point with ramification index e_i, and i be the number of a square with upper left vertex R. Let γ be the simple closed counterclockwise path in X^* around R starting at P_i. Up to homotopy we can choose γ to include P_j for every square j which contains a part of γ, thus it is made up of lifts of y^{-1}, x^{-1}, y and x (in this order). Its projection to $\pi_1(E^*, P)$ therefore is a power of $xyx^{-1}y^{-1}$.
The path γ thus starts with a path from P_i to $P_{\kappa(i)}$, then it continues to $P_{\kappa^2(i)}$ and so on. After e_i times it gets back to P_i for the first time because the path $xyx^{-1}y^{-1}$ in $\pi_1(E^*, P)$ is a simple closed path going once around the puncture of E^* and e_i is the ramification index of R. Therefore e_i is the minimal non-trivial number with $\kappa^{e_i}(i) = i$ and is hence the cycle-length of the cycle of κ containing i. □

Corollary 3.3. *For a normal origami with monodromy $f : F_2 \to S_d$ all ramification indices are equal to* $\mathrm{ord}(\kappa)$. *This holds also if f is replaced by $f : F_2 \twoheadrightarrow G$ as described in Proposition 1.6.*

Proof. For a normal origami we have $f^{-1}(\mathrm{Stab}_{S_d}(1)) = \ker(f)$ thus κ stabilizes one point if and only if it stabilizes all. Thus all cycle lengths equal $\mathrm{ord}(f(\kappa))$. The second characterization comes from identifying the image of f with the Galois group $G \cong F_2/\ker(f)$. □

Proposition 3.4. *The sum* $\sum(e_i - 1)$ *is always even.*
Conversely for $e_1, \ldots, e_n \in \mathbb{N}$ *satisfying this condition there exists an origami with ramification indices* (e_i).

Proof. The first part follows directly from the Theorem of Riemann-Hurwitz, or alternatively from the fact that the commutator κ in Prop. 3.2 is always an even permutation.
For the second part we construct an origami by concatenating horizontally the following components:

i) 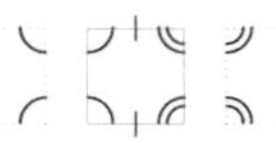

which creates a new puncture with ramification index $e = 1$,

ii) 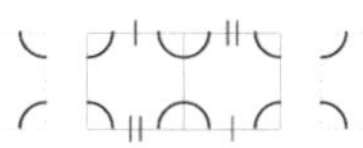

which increases the last ramification index by 2,

iii) 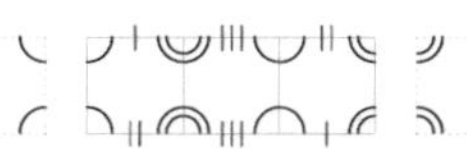

which increases the last ramification index by 1 and creates a new puncture with ramification index $e = 2$. □

3.2 Properties of the origami-curve

By Remark 2.5 we know that the normalization of the origami-curve in $\mathcal{M}_g$ is a "mirror image" of $\Gamma\backslash\mathbb{H}$. For simplicity of notation we will study geometric properties of $\Gamma\backslash\mathbb{H}$ and keep in mind that these properties coincide with the properties of our origami-curve.
Let $\boldsymbol{d}$ be the index[1] of Γ in $\mathrm{PSL}_2(\mathbb{Z})$. We get the Riemann surface $\Gamma\backslash\mathbb{H}$ by glueing $\boldsymbol{d}$ copies of a fundamental domain of $\mathrm{PSL}_2(\mathbb{Z})$ in $\mathbb{H}$. For this we use the standard fundamental triangle Δ shown in Figure 3.1, bounded by the unit circle and the lines $\mathrm{Re}(z) = \pm\frac{1}{2}$ with the point ∞i excluded (we call this point the *cusp* of Δ). Let $\left\{\Gamma g_1, \ldots, \Gamma g_{\boldsymbol{d}}\right\}$ be the right cosets of Γ. Then

[1]the usual notation would be d, as it is the degree of the map $\Gamma\backslash\mathbb{H} \to \mathrm{PSL}_2(\mathbb{Z})\backslash\mathbb{H}$, but we don't want to confuse this with the degree of the origami covering $p : X \to E$. Likewise we will use $\boldsymbol{g}$ instead of g and $\boldsymbol{n}$ instead of n when we are talking about the origami-curve.

the union of all $g_i(\Delta)$ is a fundamental domain for $\bar{\Gamma}$ (if it is connected, i.e. if every g_i is obtained from another one by multiplying S, T or their inverses from the right[2]). We identify the edges of this fundamental domain using the action of $\bar{\Gamma}$ on $\mathbb{H}$.

Example 3.5. In Example 2.6 we calculated the set $\{\text{id}, T, ST\}$ of left coset representatives of the Veech group of an origami. Its right cosets are therefore $\{\bar{\Gamma}, \bar{\Gamma}\cdot T^{-1}, \bar{\Gamma}\cdot T^{-1}S^{-1}\}$. As $S = S^{-1}$ in $\text{PSL}_2(\mathbb{Z})$ and in our case $T^2 \in \bar{\Gamma}$ we have the right cosets $\{\bar{\Gamma}, \bar{\Gamma}\cdot T, \bar{\Gamma}\cdot TS\}$. The constructed fundamental domain of the origami-curve is shown in Figure 3.1.

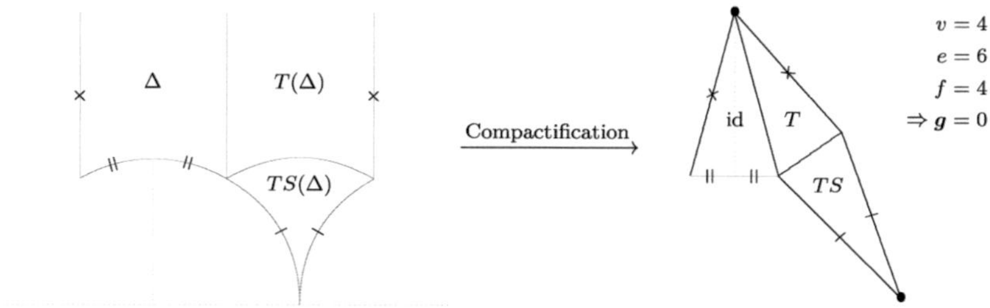

Figure 3.1: Fundamental domain of an origami-curve with two cusps (•)

Around the image of a cusp in $\bar{\Gamma}\backslash\mathbb{H}$ the copies of Δ correspond to cosets $\bar{\Gamma}g, \bar{\Gamma}gT \ldots, \bar{\Gamma}gT^{w-1}$ of $\bar{\Gamma}$ for a $g \in \bar{\Gamma}$. Thus each of those cusps corresponds to a parabolic generator gT^wg^{-1} of the projective Veech group $\bar{\Gamma}$. We call w the *width* of the cusp and $g \in \bar{\Gamma}$ a corresponding *Strebel element*.

Now let $\boldsymbol{n}$ be the number of cusps of the fundamental domain and $w_1, \ldots, w_{\boldsymbol{n}}$ the widths of the cusps. Then we have $\boldsymbol{d} = \sum w_i$. We can compactify[3] the origami-curve by adding a point at each cusp. Then our constructed fundamental domain is a partition of the origami-curve into $\boldsymbol{d}$ triangles. We would like to use this to calculate the genus $\boldsymbol{g}$ of the origami-curve, but there may be edges which are identified with itself (like for example in Figure 3.1), and then this partition would not be a triangulation. This problem can be solved by subdividing these triangles by adding an additional edge through the center of those edges. Then we can count the vertices, edges and faces of this triangulation to calculate the Euler characteristic and thus also the genus of the origami-curve.

[2]to simplify notation we will often make no difference between elements in $\text{SL}_2(\mathbb{Z})$ and their projections to $\text{PSL}_2(\mathbb{Z})$

[3]The compactification of the moduli space $\mathcal{M}_{g,n}$ will be the topic of section 3.3

Proposition 3.6. *The Veech groups of origamis which describe the same curve are conjugated. The genus $\boldsymbol{g}$ and the number of cusps $\boldsymbol{n}$ of the origami-curve, the index $\mathbf{d}$ of the Veech group Γ, the property $-1 \in \Gamma$ and the cusp widths (w_i) of an origami are all invariant under the action of* $\mathrm{SL}_2(\mathbb{Z})$.

Proof. The numbers $\boldsymbol{g}$ and $\boldsymbol{n}$ represent properties of the origami-curve and hence are obviously invariant.

Let O be an origami with Veech group Γ and $h \in \mathrm{SL}_2(\mathbb{Z})$. Then the Veech group of hO is $h\Gamma h^{-1}$ because $(h\Gamma h^{-1})hO = h\Gamma O = hO$. Therefore the Veech groups of O and hO are conjugated and hence have the same index in $\mathrm{SL}_2(\mathbb{Z})$. As -1 is in the center of $\mathrm{SL}_2(\mathbb{Z})$ we have $-1 \in \Gamma \Leftrightarrow -1 \in h\Gamma h^{-1}$.

If $\left\{\Gamma g_i\right\}$ is the system of right cosets of Γ, then $\left\{(h\Gamma h^{-1})hg_i\right\}$ is the system of right cosets of $h\Gamma h^{-1}$. Thus the constructed fundamental domain for hO will be the image of the fundamental domain for O under h. A parabolic generator $gT^w g^{-1}$ then transforms to $hgT^w g^{-1}h^{-1}$ leading to the same cusp width w. □

3.3 The boundary of moduli space

In the last section we have mentioned that we can compactify an origami-curve by adding a point at each cusp. While the origami-curve is a subset of the moduli space $\mathcal{M}_{g,n}$ the cusps do not correspond to points in this moduli space. But we can compactify $\mathcal{M}_{g.n}$ using the Deligne-Mumford-compactification $\overline{\mathcal{M}}_{g,n}$. Then the cusps of the origami-curve correspond to points in the boundary $\partial\mathcal{M}_{g,n} := \overline{\mathcal{M}}_{g,n} \setminus \mathcal{M}_{g,n}$.

The Deligne-Mumford-compactification $\overline{\mathcal{M}}_{g,n}$ is the moduli space of *stable Riemann surfaces* of genus g with n punctures[4] (c.f. [DM]). A *stable Riemann surface* is a one-dimensional compact complex space X whose only singularities are ordinary double points (i.e. points with a neighborhood isomorphic to a double cone). Additionally it is required that on each irreducible component of X of genus 0 there are at least three special points (where a point is special if it is a puncture or singular).

To each stable curve we associate the intersection graph of its irreducible components. This graph has a vertex for every irreducible component and an edge for every singularity, whose endpoints are the vertices corresponding to the components containing the singular point. We mark each vertex with a pair (g, n), where g denotes the genus of the irreducible component, and n

[4]For $g = 0$ we require $n \geq 3$, for $g = 1$ we need $n \neq 0$.

the number of marked points contained therein.

Each cusp of an origami-curve corresponds to a point in $\partial\mathcal{M}_{g,n}$, which is a stable curve with at least one singularity. The idea for the construction of this stable curve is the following: Let O be an origami defined by the covering $p: X \to E$. The Teichmüller disk Δ_O then consists of the Riemann surfaces X_A with $A \in \mathrm{SL}_2(\mathbb{R})$ together with the identity as marking $X \to X_A$. Now the ray

$$[0, \infty) \to \mathcal{T}_{g,n} \text{ defined by } t \mapsto (X_{A(t)}, \mathrm{id}) \text{ with } A(t) = \begin{pmatrix} e^t & 0 \\ 0 & e^{-t} \end{pmatrix}$$

is contained in this Teichmüller disk, and as t tends to ∞ every path on the origami in horizontal direction becomes arbitrarily small. If we think of our origami as a surface glued from squares this corresponds to contracting a horizontal line in the center of each square to a point. The thus constructed singular surface is the stable curve corresponding to the cusp of the origami-curve belonging to the triangle Δ, i.e. to the cusp with Strebel element id.

For the construction of other boundary points of the origami-curve we would have to contract closed paths in other directions. The directions in which the paths are closed are given by $s \cdot \binom{1}{0}$ for a Strebel element s and are called *Strebel directions*. But instead we can also look at the origami $s^{-1}O$, where the desired Strebel direction is transformed to the horizontal one. We therefore get

Proposition 3.7. *Let O be an origami and s a Strebel element for a cusp of its origami-curve. Then one obtains the stable curve corresponding to this cusp by contracting horizontal lines in the center of each square of the origami $s^{-1}O$.*

A formal proof can be found in [HS2], Theorem 4.1.

Example 3.8. We see in Figure 3.1 that the origami from Example 2.6 has two cusps: one with Strebel element $s_1 = \mathrm{id}$ and width $w_1 = 2$, and another one with Strebel element $s_2 = TS$ and width $w_2 = 1$. The resulting Strebel directions and the intersection graphs of the resulting stable curves are shown in Figure 3.2.

We now present an algorithm for the computation of the intersection graph. As a first step we number all closed horizontal lines which we want to contract. This is done by marking an arbitrary square of the origami with the number 1, then marking its right neighbor with the same number and so

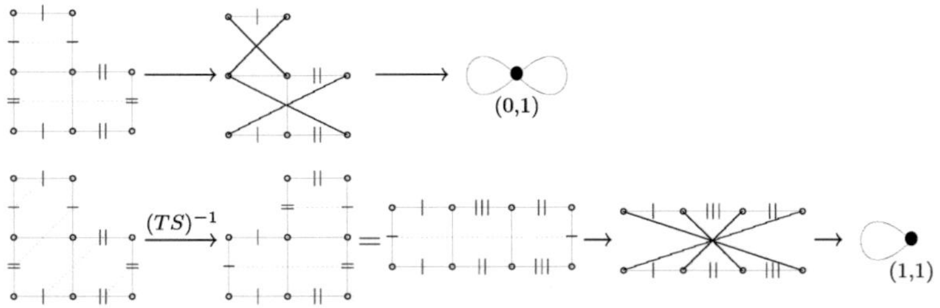

Figure 3.2: Intersection graphs of boundary points of an origami-curve

on, until we get back to the square we started with. Then we pick a new yet unmarked square and start all over for the next horizontal line with the number 2, and so on.

As second step we number the irreducible components: these are the connected areas bounded above and below by the closed horizontal lines. We mark the top half of an arbitrary square with the first number 1, then we continue marking all adjacent half squares with the same number, and continue with their neighbors. When there are no more unmarked adjacent half squares we pick a new unmarked square and start all over for the next irreducible component.

Now we can define our intersection graph: every irreducible component corresponds to a vertex in this graph, and every horizontal line to an edge. The endpoints of an edge are defined by the numbers of the components adjacent to the horizontal line.

We still have to calculate the genus g and the number n of marked points for each component. Let C be the set of bottom half squares in the component and e_i the ramification index of the marked point at the bottom left vertex of the bottom half square i. We can calculate the number of marked points by

$$n = \sum_{i \in C} \frac{1}{e_i}$$

because for every marked point R_j the e_i corresponding to R_j are equal to the ramification index e of R_j and will be counted exactly e times, thus for every marked point the sum $\sum 1/e$ is 1.

For the calculation of the genus we use the triangulation of our component which we get by contracting the horizontal lines. This triangulation has $n+d$ vertices, where d is the number of adjacent horizontal lines (i.e. the degree of the vertex in the intersection graph). The number of faces is $2\,|C|$ (for every bottom half square there is also a top half square). Therefore the Euler characteristic of the component is

$$\chi = n + d - 3|C| + 2|C| = n + d - |C|$$

and we get the genus $g = 1 - \frac{\chi}{2}$.

3.4 Automorphisms and their fixed points

The group $\mathrm{Aut}(X)$ of automorphisms of X is not invariant along the curve in moduli space. This is already obvious for the trivial origami $E \to E$. Therefore we would like to define the automorphism group of an origami to be the intersection of all the different automorphism groups occurring along this curve:

Definition 3.9. We call a bijective map $\sigma : X \to X$ an *automorphism* of the origami O, if it induces for every $A \in \mathrm{SL}_2(\mathbb{R})$ via $X \to E \to E_A$ a well-defined automorphism on E_A. The group $\mathrm{Aut}(O)$ of all such automorphisms is called the *automorphism group* of O.

The group $\mathrm{Aut}(O)$ is a subgroup of the group $\mathrm{Aut}(X)$ of automorphisms of X as a Riemann surface. Contrary to $\mathrm{Aut}(X)$ the subgroup $\mathrm{Aut}(O)$ is invariant for all complex structures, i.e. for all origamis on $c(O)$. The calculation of this group is simple: Every automorphism σ induces on E either the identity or the elliptic involution (as these are the only automorphisms of E which induce automorphisms on every E_A). Therefore the square containing P_1 has to be mapped bijectively to the square containing $\sigma(P_1)$ (either by a translation, or by a 180° rotation), and the monodromy map determines the images of the other squares.

Given an automorphism σ, the calculation of its fixed points is also simple: Let x be a fixed point of σ. If $\sigma \neq \mathrm{id}$ induces the identity on E, then no point in X^* is fixed by σ, and therefore $x \in X \setminus X^* = p^{-1}(0) = \{R_1, \dots, R_n\}$. If σ induces the elliptic involution $p(x)$ has to be fixed by this involution, therefore $x \in p^{-1}(0)$ or $x \in p^{-1}(\{(0, \frac{1}{2}), (\frac{1}{2}, 0), (\frac{1}{2}, \frac{1}{2})\}) := \{A_1, \dots, A_{3d}\}$. Thus there are at most $4d$ possible fixed points.

Every automorphism permutes the ramification points R_i and the points A_i.

Proposition 3.10. *The homomorphism*

$$\mathrm{Aut}(O) \to S_{3d} \times S_n$$

induced by the permutations of the points in $A_1, \dots A_{3d}$ and $R_1, \dots, R_n$ is (up to renumbering the points) invariant on the origami-curve.

Proof. The deformation along the origami-curve is continuous, but the points A_i and R_i are discrete in X (because their images in E are discrete). □

Example 3.11. Looking at the origami

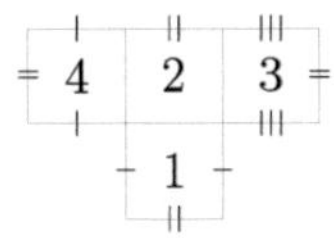

we see that there is no non-trivial automorphism which descends to the identity on E, because square 1 is horizontally adjacent to itself, but this holds for no other square, so there is no possible image for a translation of square 1 except itself. But we can rotate square 1 by 180°. Then square 2 is mapped to itself and squares 3 and 4 are exchanged:

The action of σ on the points A_i and R_i is shown in the following pictures:

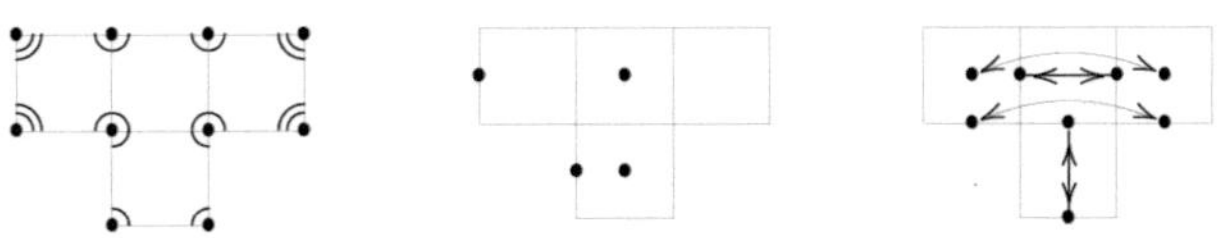

The first picture shows the two punctures R_1 and R_2 of the origami, which are both fixed. The second one shows the other four fixed points of σ, and in the third picture we see what happens to the other eight points A_i.

If an origami O is normal we can determine the group $\mathrm{Aut}(O)$ quite explicitly:

Proposition 3.12. *An origami O of degree d is normal if and only if it has d translations. In this case the group of translations is isomorphic to the Galois group G of O.*

Proof. If the origami is normal, then we can think of the squares labelled by the group elements, and for every group element right-multiplication defines a translation.

On the other hand if there are d translations this means that there are two translations σ, τ which map square number 1 to its right respectively upper neighbor. As the origami is connected they generate a group of d elements. The origami is then identical to the normal one defined by $f : F_2 \twoheadrightarrow \langle \sigma, \tau \rangle$ with $f(x) = \sigma$ and $f(y) = \tau$ (in the sense of Proposition 1.6). □

Proposition 3.13. *Let O be a normal origami with Galois group G defined by $X \to E$ and $f : F_2 \twoheadrightarrow G$. Set $\sigma := f(x)$ and $\tau := f(y)$ and let $E \to P^1(\mathbb{C}) =: P$ be the quotient map induced by the elliptic involution $z \mapsto -z$. Then the following are equivalent:*

i) $\mathrm{Aut}(O) \ncong G$,

ii) $-1 \in \Gamma(O)$,

iii) there exists some $\psi \in \mathrm{Aut}(O)$ *of order 2 not inducing the identity on* E,

iv) the composed map $X \to E \to P$ *is normal.*

v) there exists some $\varphi \in \mathrm{Aut}(G)$ *with* $\varphi(\sigma) = \sigma^{-1}$ *and* $\varphi(\tau) = \tau^{-1}$,

In this case we have

$$\mathrm{Aut}(O) \cong \mathrm{Gal}(X/P) \cong \langle G, \psi \rangle \cong C_2 \ltimes_\Phi G$$

where $\Phi : C_2 \to \mathrm{Aut}(G)$ *maps the generator of the cyclic group* C_2 *to* φ.

Proof. i) $\Leftrightarrow$ ii) is clear since automorphisms which are not translations induce the elliptic involution on E and hence have derivative -1.
i) $\Leftrightarrow$ iii) is also clear since if there is an automorphism of O which induces the elliptic involution on E we can concatenate a translation to find a non-trivial automorphism ψ which maps the square labelled with $1 \in G$ to itself. This is an automorphism of order 2.
The quotient $X/\langle G, \psi \rangle$ is isomorphic to $P = \mathbb{P}^1(\mathbb{C})$ and factors through X/G, thus iv) follows. On the other hand if $X \to P$ is normal then $G \subsetneq \mathrm{Gal}(X/P) \subseteq \mathrm{Aut}(O)$, thus iv) $\Rightarrow$ i)
The automorphism ψ maps the square σ to σ^{-1} and τ to τ^{-1}. Thus the origami defined by $f' : F_2 \to G$ with $f'(x) = \sigma^{-1}$ and $f'(y) = \tau^{-1}$ is equal to the one defined by f in $\mathrm{Aut}(G) \backslash \mathrm{Epi}(F_2, G)$, hence $f' = \varphi \circ f$ for an automorphism $\varphi \in \mathrm{Aut}(G)$. This means $\varphi(\sigma) = \sigma^{-1}$ and $\varphi(\tau) = \tau^{-1}$, therefore i) $\Rightarrow$ v). The reverse argumentation shows v) $\Rightarrow$ i).
The group of translations is isomorphic to G and contained in $\mathrm{Aut}(O)$ with index 2, hence normal. Thus we have a split exact sequence

$$0 \longrightarrow G \longrightarrow \mathrm{Aut}(O) \underset{\Phi}{\longrightarrow} C_2 \longrightarrow 0$$

Therefore $\mathrm{Aut}(O) \cong G \ltimes_\Phi C_2$. $\square$

3.5 Holomorphic differentials

An origami is naturally associated to an Abelian differential ω. Its stratum in the moduli space of Abelian differentials is determined by the multiplicities of zeroes of ω. A continuous deformation of the complex structure of an origami leads also to a continuous deformation of the Abelian differential ω. Therefore the connected component of ω in the moduli space of Abelian differentials $\mathscr{H}$ is invariant on the Teichmüller disk defined by the origami.

Kontsevich an Zorich have shown in [KZ] that each stratum contains up to 3 connected components. The connected component of ω in $\mathscr{H}$ is determined by two properties (see [KZ] 2.3, Theorem 1):

1. The property 'hyperelliptic'.
If a hyperelliptic involution[5] σ on X exists, then ω is called *hyperelliptic* if it has only one zero, or if it has two zeroes which are exchanged by σ.

2. The parity of the canonical spin structure.
A *spin structure* on X is a divisor D on X with $2D = \sum(e_i - 1)R_i$ in $\mathrm{Pic}(X)$. If all e_i are odd, then there is a *canonical spin structure* $D = \sum \frac{e_i-1}{2} R_i$. The *parity of the spin structure* D is the parity of $\ell(D) - 1 = \dim H^0(X, \mathcal{L}(D)) - 1$ (where $\mathcal{L}(D)$ is the associated invertible sheaf to D), which equals[6] the dimension the complete linear system $|D|$.

We compute both properties in Appendix A.2 for all origamis with up to eight squares. For the hyperellipticity of ω we will use the following characterization:

Proposition 3.14. *An automorphism σ of a Riemann surface X is a* hyperelliptic involution *if it has order* 2 *and* $2g + 2$ *fixed points.*

This can be proven easily using Riemann-Hurwitz. In section 3.4 we have already computed the automorphism group of an origami X (in the case where the underlying elliptic curve has only trivial automorphisms). We also have computed the fixed points of those automorphisms. This is sufficient to identify the hyperelliptic involution (if it exists) by Prop. 3.14. We can then easily determine whether ω is hyperelliptic.

For the parity of the canonical spin structure [KZ], 3.2 offers another characterization, which is more accessible for a direct computation:

[5] A *hyperelliptic involution* of X is an automorphism σ of X of order 2 such that the quotient $X/\langle\sigma\rangle$ has genus 0.

[6] For the definition of the complete linear system and the proof of the mentioned equality we refer to [Har], II.7.7 and the following definition.

Proposition 3.15. *Choose oriented smooth closed paths* (α_i, β_i) *on* X^* *representing a symplectic basis of* $H_1(X, \mathbb{Z})$. *Then the parity of the canonical spin structure of* ω *(if it exists) is equal to the parity of*

$$\varphi(\omega) := \sum_{i=1}^{g} (\mathrm{ind}_{\alpha_i} + 1)(\mathrm{ind}_{\beta_i} + 1)$$

where ind_γ *is the winding number*[7] *of a closed curve* γ.

An origami of genus g is topologically a sphere with g handles. The algorithm for surface normalization from [Sti], 1.3 can be used to transform our origami into this normalized form. A symplectic basis of $H_1(X, \mathbb{Z})$ is given there by the standard generators of the fundamental group of such a normalized surface: we get two generators for each handle, which intersect each other only once, and do not intersect any other generators. The normalized surface (together with those $2g$ generators of its fundamental group) can now be transformed back to the original origami. Thus we get the paths α_i and β_i needed for the application of the proposition.

Example 3.16. In Example 3.11 we calculated an automorphism σ of an origami of order two with six fixed points. The origami has degree $d = 4$ and $n = 2$ punctures, thus its genus is $g = 2$ and σ is a hyperelliptic involution by Prop. 3.14. There is only one zero of ω because one of the punctures has ramification index 1 and is hence not a zero of ω. Therefore the corresponding Abelian differential is contained in the hyperelliptic component.

For the calculation of the parity of the canonical spin structure we first calculate a symplectic basis of $H_1(X, \mathbb{Z})$ using the algorithm for surface normalization. The results are the paths

$$\alpha_1 = y^2, \quad \beta_1 = x, \quad \alpha_2 = x^{-1}y^{-1}x^2yx^{-2}yx, \quad \beta_2 = y^{-1}x^{-1}y^{-1}x^{-2}yx$$

which are shown[8] in Figure 3.3, where one can check easily that they form indeed a symplectic basis.

For the calculation of the winding number of a path $\gamma \in \pi_1(X^*, P)$ we first replace γ by a cyclically reduced representative of its conjugacy class (this corresponds to selecting another basepoint P for $\pi_1(X^*, P)$). Then at each point in the same fiber as P over E^* the path either turns 90° left or 90° right or it continues in its previous direction. We can thus compute the winding

[7]Note that the winding numbers depend on the translation structure on X given by ω.

[8]Recall that we write elements of the fundamental group from right to left, as stated in the footnote on page 2

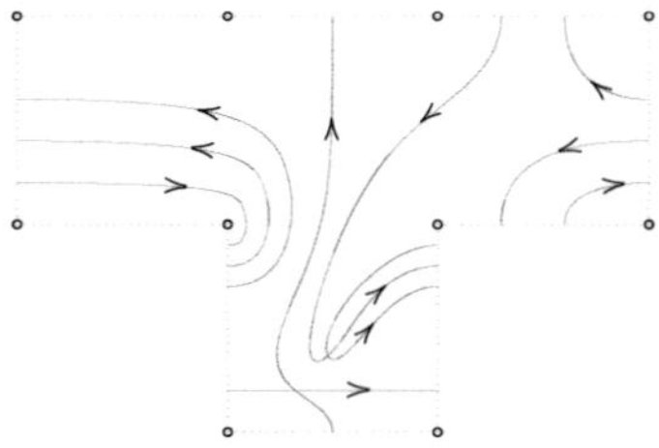

Figure 3.3: Representatives of a symplectic basis of $H_1(X, \mathbb{Z})$.

number ind_γ easily by counting all the left-turns, subtracting the right-turns, and dividing the result by 4.

For our example we get

$$\text{ind}_{\alpha_1} = \text{ind}_{\beta_1} = \text{ind}_{\alpha_2} = 0, \quad \text{ind}_{\beta_2} = \tfrac{1}{4}(5-1) = 1$$

Therefore the parity of the canonical spin structure in this case is the parity of $1 \cdot 1 + 1 \cdot 2 = 3$, which is odd.

3.6 The Galois group

Let O and $\overline{O}$ be two origamis defined by $p : X \to E$ and $\overline{p} : \overline{X} \to E$ respectively. We say that $\overline{O}$ is a *covering* of O if there is a covering $h : \overline{X} \to X$ such that $\overline{p} = p \circ h$. Such a covering induces an inclusion of the corresponding fundamental groups:

$$\overline{H} = \pi_1(\overline{X}^*, \overline{P}) \hookrightarrow \pi_1(X^*, P) = H$$

Conversely, if we have two finite index subgroups H and $\overline{H}$ of F_2 with $\overline{H} \subset H$, then the origami defined by $\overline{H}$ covers the one defined by H.

Proposition 3.17. *Every origami O has a minimal normal covering O', i.e. a covering $O' \to O$ such that O' is a normal origami and every other covering of a normal origami $\overline{O}$ over O also covers O'.*

Proof. Let H be the fundamental group of O as above and $\overline{H}$ the fundamental group of a normal covering of O. Then $\overline{H}$ is a normal subgroup of F_2 contained in H. Hence

$$\overline{H} \subseteq N := \bigcap_{\gamma \in F_2} \gamma^{-1} H \gamma$$

where N is a normal subgroup of F_2. Hence the origami O' defined by N is a normal covering of O, and the stated inclusion proves that $\overline{O}$ covers O'. □

Proposition 3.18. *Let O be an origami with monodromy $f : F_2 \to S_d$. Then the Galois group G of the minimal normal covering of O is isomorphic to the image of f and this isomorphy type is invariant under the action of* $\mathrm{SL}_2(\mathbb{Z})$. *We call this group the* Galois group *of O.*

Proof. With the notation as in Prop. 3.17 we use $H = f^{-1}(\mathrm{Stab}_{S_d}(1))$ to get the equation

$$\begin{aligned}\ker(f) = f^{-1}(\mathrm{id}) &= f^{-1}\Big(\bigcap_{i\in\{1,\dots,d\}} \mathrm{Stab}_{S_d}(i)\Big)\\ &= \bigcap_{\gamma\in F_2} \gamma^{-1}f^{-1}(\mathrm{Stab}_{S_d}(1))\gamma = \bigcap_{\gamma\in F_2} \gamma^{-1}H\gamma = N\end{aligned}$$

Therefore we have $f(F_2) \cong F_2/\ker(f) = F_2/N \cong G$.
Obviously $f(F_2) = (f \circ \varphi)(F_2)$ for every $\varphi \in \mathrm{Aut}(F_2)$, hence the invariance under the action of $\mathrm{SL}_2(\mathbb{Z})$. □

Example 3.19. Let O be the origami defined by $\sigma_x = (1\,2\,3)(4\,5\,6)$ and $\sigma_y = (2\,4)(3\,5)$. The monodromy map $f : F_2 \to S_6$ maps x to σ_x and y to σ_y, therefore the image of f is the subgroup of S_6 generated by σ_x and σ_y. This group is isomorphic to A_4. Thus A_4 is the Galois group of O. The corresponding normal covering is shown in Figure 3.4.

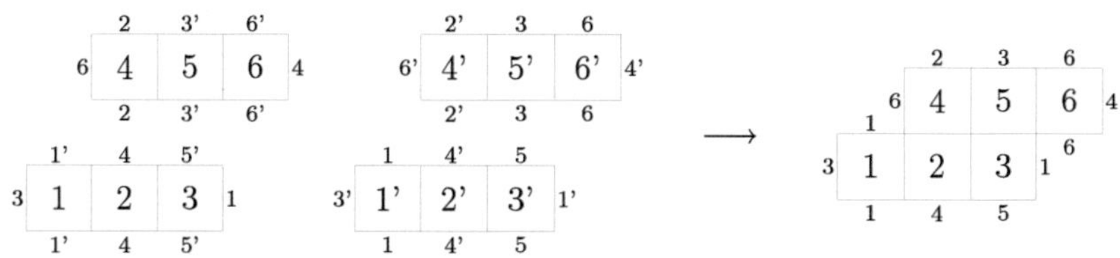

Figure 3.4: Minimal normal covering of an origami

Chapter 4

The non-archimedean world

Up to now we have only considered origamis as compact Riemann surfaces, i.e. projective nonsingular curves defined over $\mathbb{C}$. In this chapter we want to make an analogous definition for curves defined over the completed algebraic closure $\mathbb{C}_p$ of the field of p-adic numbers. As a p-adic analogue of Riemann surfaces we use *Mumford curves*. Mumford showed in [Mum] that these curves can be uniformized as Ω/G where $G \subset \mathrm{PGL}_2(\mathbb{C}_p)$ acts[1] properly discontinuously on $\Omega \subset \mathbb{P}^1$. We are going to use this property as a definition. Contrary to the complex setting not every projective nonsingular curve is a Mumford curve. For more details about Mumford curves we refer to [GP].

We will define a p-adic origami as a covering $X \to E$ of Mumford curves defined over $\mathbb{C}_p$ with at most one branch point, where $g(E) = 1$ (i.e. E is a Tate curve). These p-adic origamis also form curves in the moduli space of curves over $\mathbb{C}_p$, which are connected to corresponding origami curves over $\mathbb{C}$. This connection between origamis over $\mathbb{C}_p$ and over $\mathbb{C}$ will be the topic of Chapter 5.

In the present chapter we will propose methods for the calculation of some of the invariants introduced in Chapter 3. We restrict our considerations to normal origamis with non-trivial ramification (i.e. non-abelian Galois group). Those origamis will be classified in Section 4.3. It is already known that every finite group occurs as a Galois group of a covering of a Mumford curve over an elliptic curve, which was proven in [PV], Theorem 1.2. Hence the interesting part will be to control the number of branch points of such coverings. To do this we will use some results of Kato and Bradley on Mumford orbifolds.

[1]For better readability we will continue to write this quotient space as Ω/G even though the action of G is a left action.

4.1 Discontinuous groups

After defining Mumford curves we will construct in this section the *Bruhat-Tits-Tree* $\mathcal{B}$ for an extension of $\mathbb{Q}_p$ using a quite concrete definition from [Her]. A Mumford curve is closely related to the quotient graph of an action of G on a subtree of $\mathcal{B}$. Often one defines a suitable subtree such that the quotient becomes a finite graph, but instead we will follow Kato [Kat2], who uses a slightly larger quotient graph, which can be used to control the ramification behavior of coverings of Mumford curves.

Definition 4.1. Let k be a field which is complete with respect to a non-archimedean valuation and G a subgroup of $\mathrm{PGL}_2(k)$. A point $x \in \mathbb{P}^1(k)$ is called a *limit point* of G, if there exist pairwise different $\gamma_n \in G$ $(n \in \mathbb{N})$ and a point $y \in \mathbb{P}^1(k)$ satisfying $\lim \gamma_n(y) = x$. The set of limit points is denoted by $\mathcal{L}(G)$.

G is a *discontinuous group*, if $\Omega(G) := \mathbb{P}^1(k) \setminus \mathcal{L}(G)$ is nonempty and for each $x \in \mathbb{P}^1(k)$ the closure $\overline{Gx}$ of its orbit is compact. A discontinuous group G is called a *Schottky group* if it is finitely generated and has no non-trivial elements of finite order. Every Schottky group is free ([GP], Theorem I.3.1).

A discontinuous group G acts properly discontinuously on $\Omega(G)$. For a Schottky group G we know from [GP], Theorem III.2.2. that the quotient $\Omega(G)/G$ is the analytification of an algebraic curve. Such a curve is called a *Mumford curve*.

If an arbitrary group $G \subset \mathrm{PGL}_2(k)$ contains a discontinuous group G' of finite index, then $\mathcal{L}(G) = \mathcal{L}(G')$ and G is also discontinuous. We know from [GP], Ch. I, Theorem 3.1 that every finitely generated discontinuous group contains a Schottky group as a subgroup of finite index.

Let $k \subset \mathbb{C}_p$ be a finitely generated extension of $\mathbb{Q}_p$. Then the set of absolute values $|k^\times| := \{|x| : x \in k^\times\}$ is a discrete set in $\mathbb{R}^\times$. For $r \in |k^\times|$ and $x \in k$ let $B(x,r) := \{y \in k : |x-y| \leq r\}$ be the "closed" ball[2] around x. Construct a graph with vertices $B(x,r)$ and insert edges connecting $B(x,r)$ and $B(x',r')$ with $B(x,r) \subset B(x',r')$ and $[r,r'] \cap |k^\times| = \{r,r'\}$. This graph is a simplicial tree, called the *Bruhat-Tits-Tree* $\mathcal{B}(k)$. The ends[3] of this graph correspond bijectively to the points in $\mathbb{P}^1(\hat{k})$, where $\hat{k}$ denotes the completion of k. The action of $\mathrm{PGL}_2(k)$ on $\mathbb{P}^1(k)$ can be continued to an action on $\mathcal{B}(k)$, and we can modify $\mathcal{B}(k)$ by adding vertices such that this action is without inversion.

[2]Note that as $|k^\times|$ is discrete the ball $B(x,r)$ is both open and closed for the topology induced by the p-adic norm.

[3]An *end of a graph* is an infinite ray up to finitely many edges.

Let $\gamma \in \mathrm{PGL}_2(k)$ be hyperbolic or elliptic with two fixed points in $\mathbb{P}^1(k)$. In this case we define the *axis* $\mathcal{A}(\gamma)$ to be the infinite path connecting the two ends of $\mathcal{B}(k)$ corresponding to the fixed points of γ. A hyperbolic element $\gamma \in \mathrm{PGL}_2$ acts on $\mathcal{A}(\gamma)$ by shifting the whole axis towards the end corresponding to the attracting fixpoint of γ. An elliptic element fixes $\mathcal{A}(\gamma)$ pointwise. It has additional fixed points in $\mathcal{B}(k)$ if and only if $\mathrm{ord}(\gamma)$ is a power of p (this is made more precise in [Her], Lemma 3).

Let G be a finitely generated discontinuous subgroup of $\mathrm{PGL}_2(\mathbb{C}_p)$ and let $F(\gamma)$ be the set of the two fixed points of $\gamma \in G$ in $\mathbb{P}^1(\mathbb{C}_p)$. Now let k be the extension of $\mathbb{Q}_p$ generated by the coefficients of the generators of G and the fixed points of representatives of every conjugacy class of elliptic elements in G. There are only finitely many such conjugacy classes, hence k is a finitely generated field extension of $\mathbb{Q}_p$ and we can therefore construct the Bruhat-Tits-Tree $\mathcal{B}(k)$ as described above. Note that by construction $G \subset \mathrm{PGL}_2(k)$ and $F(\gamma) \subset \mathbb{P}^1(k)$ for all elliptic elements $\gamma \in G$. Moreover we have $F(\gamma) \subset \mathbb{P}^1(\hat{k})$ for every hyperbolic element $\gamma \in G$ because the endpoints of $\mathcal{A}(\gamma)$ correspond to the fixed points of γ.

As G is discontinuous, G contains only hyperbolic and elliptic elements ([Kat2], Lemma 4.2). The set of all fixed points of G

$$F(G) := \bigcup_{\gamma \in G} F(\gamma)$$

is a G-invariant subset of $\mathbb{P}^1(\hat{k})$, therefore G also acts on the subtree $\mathcal{T}^*(G)$ of $\mathcal{B}(k)$ generated by the ends corresponding to $F(G)$. We can now construct the quotient graph $\mathcal{G}^*(G) := \mathcal{T}^*(G)/G$. Each axis of a hyperbolic element will be mapped to a circle in $\mathcal{G}^*(G)$, while each end of $\mathcal{T}^*(G)$ corresponding to a fixed point of an elliptic element but not to a fixed point of a hyperbolic element will be mapped to an end of $\mathcal{G}^*(G)$. We now turn $\mathcal{G}^*(G)$ into a graph of groups[4] by labeling the image of a vertex resp. edge $x \in \mathcal{T}^*(G)$ with the conjugacy class of the stabilizing group G_x of x.

The graph $\mathcal{G}^*(G)$ contains a lot of useful information about the Mumford curve $\Omega(G)/G$. The ramification points of the covering $\Omega(G) \to \Omega(G)/G$ are the fixed points of elliptic elements of G which are not fixed points of hyperbolic elements ([Kat2], Prop. 5.6.2). Therefore the branch points correspond bijectively to the ends of $\mathcal{G}^*(G)$. The stabilizing group of such an end is a cyclic group whose order equals the corresponding ramification index. And by studying the action of the hyperbolic elements, we find that the genus[5] of

[4]In this chapter a *graph* will always mean a *graph of groups* to shorten notation. For the definition of a *graph of groups* we refer to [Ser], I.4.4.

[5]By the *genus of a graph* we mean its first Betti number.

$\mathcal{G}^*(G)$ equals the genus of $\Omega(G)/G$ ([Kat2], §5.6.0).

Definition 4.2. We define a *p-adic origami* to be a covering of Mumford curves $X \to E$ ramified above at most one point with $g(E) = 1$.

Starting from a Mumford curve $X = \Omega/\Gamma$ for a Schottky group Γ we will later consider the quotient map to $E = \Omega/G$ for an extension G of Γ. As Γ is free the map $\Omega \to \Omega/\Gamma$ is unramified, therefore the branch points of $X \to E$ are equal to those of $\Omega \to \Omega/G$. Thus both necessary informations for $X \to E$ to be a p-adic origami (the genus of E and the number of branch points) are coded in the quotient graph $\mathcal{G}^*(G)$ (as its genus and the number of its ends). We will now give two examples of how we can use this to construct p-adic origamis.

Example 4.3. Let $p > 5$ and $n \in \mathbb{N}$ be odd, $\zeta \in \mathbb{C}_p$ be a primitive n-th root of unity, $q \in \mathbb{C}_p$ with $|q| < |1 - \zeta|$ and set

$$\delta = \begin{pmatrix} \zeta & 0 \\ 0 & 1 \end{pmatrix}; \quad \sigma = \begin{pmatrix} 0 & 1 \\ 1 & 0 \end{pmatrix}; \quad \gamma = \begin{pmatrix} 1+q & 1-q \\ 1-q & 1+q \end{pmatrix}$$

Thus δ is elliptic of order n with fixed points 0 and ∞, the involution σ exchanges the fixed points of δ and has fixed points 1 and -1, and γ is hyperbolic with the same fixed points as σ. Then we have

- $\gamma\sigma = \sigma\gamma$ and $\delta\sigma = \sigma\delta^{-1}$
- $\Delta := \langle \delta, \sigma \rangle$ is the dihedral group D_n and fixes a single vertex $\mathcal{A}(\delta) \cap \mathcal{A}(\sigma)$.
- $\Gamma := \langle \delta^i \gamma \delta^{-i} : i \in \{0, \ldots, n-1\} \rangle$ is a Schottky group on n free generators.
- Γ is a normal subgroup of $G := \langle \delta, \sigma, \gamma \rangle$ of index $2n$, hence $\Omega(G) = \Omega(\Gamma) =: \Omega$. It is the kernel of the map $\varphi : G \to \Delta$ defined by $\varphi|_\Delta = \mathrm{id}$ and $\varphi(\gamma) = 1$.
- The quotient graph $\mathcal{G}^*$ of Ω/G is

  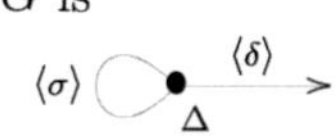

  where we use the arrow to indicate an end of this graph.
- Since $\mathcal{G}^*$ has genus 1 and one end the map $\Omega/\Gamma \to \Omega/G$ is a normal p-adic origami with Galois group $G/\Gamma \cong D_n$.

A more detailed investigation of this origami can be found in [Kre], Bemerkung 4.3.

Example 4.4. Let $p > 5$ and $\zeta \in \mathbb{C}_p$ be a third root of unity, $q \in \mathbb{C}_p$ with $|q|$ small enough[6] and set

$$\delta = \begin{pmatrix} \zeta & 0 \\ 0 & 1 \end{pmatrix}; \quad \sigma = \begin{pmatrix} -1 & 1 \\ 2 & 1 \end{pmatrix}; \quad \gamma = \begin{pmatrix} q & 0 \\ 0 & 1 \end{pmatrix}$$

Thus δ is elliptic of order 3 with fixed points 0 and ∞, and γ is hyperbolic with the same fixed points. The fixed points of the involution σ are $-\frac{1}{2}(1 \pm \sqrt{3})$, those of $\sigma\delta\sigma$ are $\sigma(0) = 1$ and $\sigma(\infty) = -\frac{1}{2}$. Then we have

- $\gamma\delta = \delta\gamma$ and $(\delta\sigma)^3 = \mathrm{id}$
- $\Delta := \langle \delta, \sigma \rangle$ is the tetrahedral group A_4 and fixes a single vertex $\mathcal{A}(\delta) \cap \mathcal{A}(\sigma)$.
- $\Gamma := \langle \alpha\gamma\alpha^{-1} : \alpha \in T \rangle$ is a Schottky group on 4 free generators.
- Γ is a normal subgroup of $G := \langle \delta, \sigma, \gamma \rangle$ of index 12, hence $\Omega(G) = \Omega(\Gamma) =: \Omega$. It is the kernel of the map $\varphi : G \to \Delta$ defined by $\varphi|_\Delta = \mathrm{id}$ and $\varphi(\gamma) = 1$.
- The quotient graph $\mathcal{G}^*$ of Ω/G is

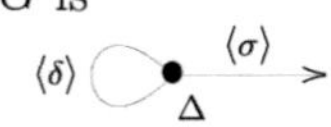

- Since $\mathcal{G}^*$ has genus 1 and one end the map $\Omega(\Gamma)/\Gamma \to \Omega(G)/G$ is a normal p-adic origami with Galois group $G/\Gamma \cong A_4$.

A more detailed investigation of this origami can be found in [Kre], Bemerkung 4.4.

We will see in Section 4.3 that the quotient graphs $\mathcal{G}^*$ of all non-trivial normal p-adic origamis look similar. We will then use this to investigate how the groups G and Γ have to be chosen such that the map $\Omega/\Gamma \to \Omega/G$ becomes a p-adic origami. But first we have to study the quotient graph $\mathcal{G}^*$ more closely.

4.2 Properties of the quotient graph

A graph of groups $\mathcal{G}^*$ is called *p-realizable*[7], if there exists a finitely generated discontinuous group $G \subset \mathrm{PGL}_2(\mathbb{C}_p)$ with $\mathcal{G}^* = \mathcal{G}^*(G)$. Let $\mathcal{G}^{\mathrm{nt}}$ resp. $\mathcal{G}^{\mathrm{nc}}$ be the subgraph of $\mathcal{G}^*$ containing only vertices and edges with non-trivial resp. non-cyclic groups.

[6] This is made more precise in [Kre], Bemerkung 4.4.

[7] In this section we will often just write *realizable* if the statements hold for arbitrary p.

Theorem 4.5. *The number of ends of a realizable graph $\mathcal{G}^*$ is*

$$n = \chi(\mathcal{G}^{\mathrm{nc}}) + 2\chi(\mathcal{G}^{\mathrm{nt}})$$

where $\chi(\mathcal{G})$ is the Euler-characteristic[8] *of a graph $\mathcal{G}$ (for infinite $\mathcal{G}$ we take the limit of χ for all finite subgraphs of $\mathcal{G}$).*

Proof. Let D resp. d be the the number of vertices resp. edges in $\mathcal{G}^{\mathrm{nc}}$. Then $\chi(\mathcal{G}^{\mathrm{nc}}) = D - d$. Analogously let C resp. c be the number of vertices resp. edges in $\mathcal{G}^{\mathrm{nt}} \setminus \mathcal{G}^{\mathrm{nc}}$. Thus $\chi(\mathcal{G}^{\mathrm{nt}}) = (C + D) - (c + d)$. Then we have to show

$$n = D - d + 2((C + D) - (c + d)) = 2(C - c) + 3(D - d)$$

Thus our statement is just a reformulation of [Bra2], Theorem 1. □

Lemma 4.6. *Let G be a finitely generated discontinuous group and let $\mathcal{N}$ be a subgraph of $\mathcal{G}^*(G)$. Then there exists a subgroup N of G with quotient graph $\mathcal{N}^* := \mathcal{G}^*(N) \supset \mathcal{N}$, such that the difference between the two graphs $\mathcal{N}$ and $\mathcal{N}^*$ is contractible*[9]*, except for the ends of $\mathcal{N}^*$.*

Proof. Choose a spanning tree of $\mathcal{N}$ by deleting edges $\{e_1, \ldots, e_g\}$ and let $\hat{\mathcal{N}}$ be a preimage of this spanning tree in $\mathcal{T}^*(G)$. For each edge e_i connecting vertices v_i and w_i let $\hat{e}_i$ be the lift of e_i with $\hat{v}_i \in \hat{\mathcal{N}}$ and $\hat{e}_i'$ the lift with $\hat{w}_i' \in \hat{\mathcal{N}}$. The other endpoints $\hat{w}_i$ and $\hat{v}_i'$ of $\hat{e}_i$ resp. $\hat{e}_i'$ cannot be contained in $\hat{\mathcal{N}}$ because otherwise $\hat{\mathcal{N}}$ would contain a circle. Let N be the subgroup of G generated by all stabilizers of vertices in $\hat{\mathcal{N}}$ and for each edge e_i a hyperbolic element γ_i mapping an $\hat{v}_i$ to $\hat{v}_i'$.

Thus N is isomorphic to the fundamental group[10] of the graph of groups $\mathcal{N}$ by [Ser], I.5.4, Theorem 13. The stabilizers in $\mathcal{N}$ do not change if we restrict the action from G to N, neither do the identifications of vertices via the γ_i, hence the quotient graph $\mathcal{T}^*(G)/N$ contains $\mathcal{N}$. Both graphs have the common fundamental group N, thus their difference cannot change their fundamental group, and therefore has to be contractible.

The tree $\mathcal{T}^*(N)$ is contained in $\mathcal{T}^*(G)$, hence $\mathcal{G}^*(N) = \mathcal{T}^*(N)/N$ is contained in $T^*(G)/N$. Again both graphs have the common fundamental group N and hence their difference is contractible. □

[8] Recall that the *Euler-characteristic* (number of vertices minus number of edges) equals the difference of the first two Betti-numbers (number of connected components minus genus).

[9] An edge in a graph of groups may be *contracted* if it is not a loop and the inclusion of the edge group into one of its vertex groups is an isomorphism. After the contraction only the other vertex remains. Such a contraction does not change the fundamental group[10] of the graph.

[10] For the definition of the *fundamental group* of a graph of groups we refer to [Ser], I.5.1.

Proposition 4.7. *Let $\mathcal{G}^*$ be a realizable graph, and let $\mathcal{C}$ be a connected component of $\mathcal{G}^{\mathrm{nc}}$. Then there exists a realizable graph $\mathcal{N}^*$ with $\mathcal{N}^{\mathrm{nc}} = \mathcal{C}$ (up to contractions) and $g(\mathcal{N}^{\mathrm{nc}}) = g(\mathcal{N}^*)$.*

Proof. Let G be a finitely generated discontinuous group with $\mathcal{G}^*(G) = \mathcal{G}^*$. Subdivide all edges emanating from $\mathcal{C}$ (which all have cyclic stabilizers), and let $\partial(\mathcal{C})$ be the set of all resulting edges in $\mathcal{G}^* \setminus \mathcal{C}$ which still have a common vertex with $\mathcal{C}$. For the graph $\mathcal{C} \cup \partial(\mathcal{C})$ Lemma 4.6 yields a graph $\mathcal{N}^*$ with $\mathcal{N}^* \supset \mathcal{C}$. As $\mathcal{N}^*$ and $\mathcal{C}$ differ up to contraction only by ends, and ends are stabilized by cyclic groups, we get $\mathcal{N}^{\mathrm{nc}} = \mathcal{C}$ (up to contractions) and $g(\mathcal{N}^{\mathrm{nc}}) = g(\mathcal{C}) = g(\mathcal{N}^*)$. □

Proposition 4.8. *Let $\mathcal{G}$ be a connected graph of noncyclic groups with $g(\mathcal{G}) > 0$. Then there exists no realizable graph $\mathcal{G}^*$ with $\mathcal{G}^{\mathrm{nc}} = \mathcal{G}$ (up to contractions) and $g(\mathcal{G}^{\mathrm{nc}}) = g(\mathcal{G}^*)$.*

Proof. Assume there is a finitely generated discontinuous group G such that $\mathcal{G}^* := \mathcal{G}^*(G)$ has the stated properties. $\mathcal{G}^{\mathrm{nt}}$ is connected, because if σ and τ are elements of stabilizers of two different connected components of $\mathcal{G}^{\mathrm{nt}}$, then $\sigma\tau$ is hyperbolic and its axis contains the path p between the axes of σ and τ. The image of $\mathcal{A}(\sigma\tau)$ in $\mathcal{G}^*$ is a circle which contains p and hence an edge with trivial stabilizer. This would imply $g(\mathcal{G}^*) > g(\mathcal{G}^{\mathrm{nc}})$ contrary to the assumption.

Thus we have $g(\mathcal{G}^{\mathrm{nt}}) = g(\mathcal{G}^{\mathrm{nc}})$ and both $\mathcal{G}^{\mathrm{nt}}$ and $\mathcal{G}^{\mathrm{nc}}$ are connected, hence $\chi(\mathcal{G}^{\mathrm{nt}}) = \chi(\mathcal{G}^{\mathrm{nc}})$. Then Theorem 4.5 states $\chi(\mathcal{G}^{\mathrm{nt}}) \geq 0$. But we have $\chi(\mathcal{G}^{\mathrm{nt}}) < 1$ because $\mathcal{G}^{\mathrm{nt}}$ is connected and has positive genus. We thus get $\chi(\mathcal{G}^{\mathrm{nt}}) = \chi(\mathcal{G}^{\mathrm{nc}}) = 0$ and therefore $g(\mathcal{G}^{\mathrm{nt}}) = g(\mathcal{G}^{\mathrm{nc}}) = 1$. Hence by Theorem 4.5 the graph $\mathcal{G}^*$ has no ends.

G contains a normal subgroup Γ of finite index which is a Schottky group. As $\mathcal{G}^*$ has no ends, the covering $\Omega(G)/\Gamma \to \Omega(G)/G$ is unramified. As $g(\Omega(G)/G) = 1$, we conclude $g(\Omega(G)/\Gamma) = 1$ by Riemann-Hurwitz. Therefore Γ is generated by a single hyperbolic element γ. All the elements in G have the same axis as γ (because otherwise there would be ramification points). Therefore every finite subgroup of G is cyclic, which contradicts the assumption. □

Proposition 4.9. *Let $\mathcal{G}^*$ be a realizable graph. Then $g(\mathcal{G}^{\mathrm{nc}}) = 0$.*

Proof. For every connected component of $\mathcal{G}^{\mathrm{nc}}$ this follows from Propositions 4.7 and 4.8. □

Definition 4.10. Let $G \subset \mathrm{PGL}_2(\mathbb{C}_p)$ be a discontinuous group, $g(\Omega(G)/G) = 0$ and $\Omega(G) \to \Omega(G)/G$ ramified over exactly three points with ramification indices n_1, n_2, n_3. Then we call G a (p-adic) *triangle group* of type $\Delta(n_1, n_2, n_3)$.

The graph $\mathcal{G}^*(\Delta(n_1, n_2, n_3))$ is a tree with exactly three ends, corresponding to the three branch points. Conversely if G is a discontinuous group, $\mathcal{G}^{\mathrm{nt}}$ is a tree, and $\mathcal{G}^{\mathrm{nc}}$ is connected, then $\chi(\mathcal{G}^{\mathrm{nt}}) = \chi(\mathcal{G}^{\mathrm{nc}}) = 1$, and G is a triangle group by Theorem 4.5.

Theorem 4.11. *Let $\mathcal{G}^*$ be a realizable graph and $\mathcal{C}$ be a connected component of $\mathcal{G}^{\mathrm{nc}}$. Then the fundamental group of $\mathcal{C}$ is a triangle group Δ. This means that $\mathcal{C}$ can be replaced by a single vertex with vertex group Δ without changing the fundamental group of $\mathcal{G}^*$.*

Proof. Let $\mathcal{N}^*$ be the realizable graph associated to $\mathcal{C}$ by Proposition 4.7. By Proposition 4.9 this graph has genus zero, so by Theorem 4.5 it has three ends. Therefore the discontinuous group Δ with quotient graph $\mathcal{N}^*$ is a triangle group. □

Now we know that a p-realizable graph $\mathcal{G}$ with $\mathcal{G}^{\mathrm{nc}} \neq \emptyset$ can be made up of vertices with p-adic triangle groups connected by edges with cyclic groups, it becomes vitally important to find all triangle groups which can occur. Fortunately for $p > 5$ those triangle groups are well-known:

Theorem 4.12. *For every p there exist the classical spherical triangle groups (i.e. those with $\frac{1}{n_1} + \frac{1}{n_2} + \frac{1}{n_3} > 1$): the dihedral group $D_n = \Delta(2, 2, n)$, and the symmetry groups of the platonic solids $A_4 = \Delta(2, 3, 3)$, $S_4 = \Delta(2, 3, 4)$ and $A_5 = \Delta(2, 3, 5)$. For $p > 5$ there are no other p-adic triangle groups.*

Proof. Let Δ be one of the given groups. It is a finite subgroup of $\mathrm{PGL}_2(\mathbb{C}_p)$ and hence discontinuous. Its quotient graph $\mathcal{G}^*(\Delta)$ consists up to contraction of only one vertex (otherwise Δ would be a non-trivial amalgam or HNN-extension of smaller groups, hence would not be finite). This vertex has to be fixed by the whole group, which is non-cyclic, hence $\chi(\mathcal{G}^{\mathrm{nc}}) = \chi(\mathcal{G}^{\mathrm{nt}}) = 1$. Therefore Δ is a triangle group by Theorem 4.5.

Now let Δ be a triangle group for $p > 5$. We have $g(\mathcal{G}^*(\Delta)) = 0$, hence $\chi(\mathcal{G}^{\mathrm{nc}}) \geq 1$ and $\chi(\mathcal{G}^{\mathrm{nt}}) \geq 1$. By Theorem 4.5 we have then $\chi(\mathcal{G}^{\mathrm{nc}}) = \chi(\mathcal{G}^{\mathrm{nt}}) = 1$, hence both graphs are connected. One can show that for $p > 5$ all edges of a realizable tree of groups can be contracted, which leaves a single vertex. The stabilizer of a vertex always is a finite subgroup of $\mathrm{PGL}_2(\mathbb{C}_p)$ and hence either cyclic or isomorphic to one of the groups stated (for a proof we refer to [Kre], Satz 2.7). □

For $p \leq 5$ there are additional non-spherical triangle groups. Bradley, Kato and Voskuil are currently working on their classification [BKV]. A preliminary version and an idea of the proofs can be found in [Kat1].

Example 4.13. For $p = 5$ an elliptic element δ of order 5 fixes not only its axis, but also all vertices contained in a small tube around this axis. Thus if we start with a vertex $v \in \mathcal{B}$ on the axis of δ fixed by a dihedral group D_5 generated by an element σ of order 2 and δ, then δ fixes also other vertices on the axis of σ. These vertices have the stabilizer $\langle \sigma, \delta \rangle \cong D_5$ and we can find two elements τ and τ' of order 3, each with an axis through one of those vertices but not through v, such that the stabilizers of these two vertices under the action of $G := \langle \sigma, \delta, \tau, \tau' \rangle$ are $\langle \sigma, \delta, \tau \rangle \cong A_5$ and $\langle \sigma, \delta, \tau' \rangle \cong A_5$ respectively. If we do all this carefully we can get a discontinuous group whose quotient graph looks like this:

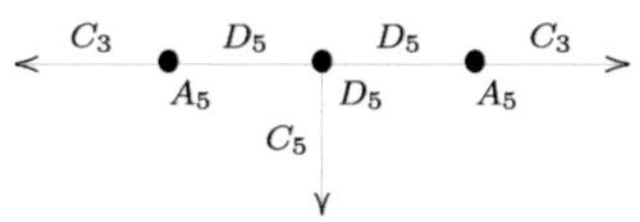

The generated group is thus a p-adic triangle group of type $\Delta(3, 3, 5)$. It is the fundamental group of the graph shown above, which is $A_5 *_{D_5} A_5$, where $*_{D_5}$ is the amalgamated product over the common subgroup D_5. Details about amalgams as fundamental groups of trees of groups can be found in [Ser], I.4.5.

One can generalize this example by starting with the dihedral group D_{5n} for $n \in \mathbb{N}$. Then δ^n has order 5 and can be used to construct two stabilizers isomorphic to A_5. This results in a discontinuous group $A_5 *_{D_5} D_{5n} *_{D_5} A_5$, which is a p-adic triangle group of type $\Delta(3, 3, 5n)$.

4.3 Normal p-adic origamis

After the preliminaries in the last two sections we are now ready to formulate our main result. We will restrict ourselves to ramified p-adic origamis, i.e. the case $g(X) > 1$. We are particularly interested in normal origamis, which we will now classify:

Theorem 4.14. *Let $X \to E$ be a normal p-adic origami with $g(X) > 1$. Then there a is discontinuous group G and a Schottky group $\Gamma \lhd G$ of finite index such that $X \cong \Omega/\Gamma$ and $E \cong \Omega/G$ with $\Omega := \Omega(\Gamma) = \Omega(G)$.*

The group G is isomorphic to the fundamental group of the graph of groups

where Δ is a p-adic triangle group of type $\Delta(a,a,b)$. This means that G is isomorphic to the fundamental group of this graph.
Thus we get

$$G \cong \langle \Delta, \gamma;\ \gamma\alpha_1 = \alpha_2\gamma \rangle \ \textit{with}\ \alpha_i \in \Delta \ \textit{of order}\ a.$$

The Galois group of the origami is G/Γ.

Proof. X is a Mumford curve, hence there is a Schottky group $\Gamma \subset \mathrm{PGL}_2(\mathbb{C}_p)$ such that $X \cong \Omega(\Gamma)/\Gamma$. The automorphism group $\operatorname{Aut} X$ is isomorphic to N/Γ, where N is the normalizer of Γ in $\mathrm{PGL}_2(\mathbb{C}_p)$ (this is a theorem from [GP], VII.2). The Galois group of the covering $X \to E$ is a finite subgroup of $\operatorname{Aut} X$ and therefore takes the form G/Γ, where Γ is a normal subgroup in $G \subseteq N$ of finite index. In this case G is discontinuous and $\Omega(G) = \Omega(\Gamma)$.
The genus of $\mathcal{G}^* := \mathcal{G}^*(G)$ equals the genus of E, which is 1. The number of ends of $\mathcal{G}^*$ equals the number of branch points of the map $\Omega \to \Omega/G$. As the map $\Omega \to \Omega/\Gamma$ is unramified, this number equals the number of branch points of $X \cong \Omega/\Gamma \to E \cong \Omega/G$, which is also 1. Thus $\mathcal{G}^*$ is a realizable graph of genus one with one end. The stabilizer of this end is a cyclic group whose order equals the ramification index above the branch point.

We now prove that $\mathcal{G}^{\mathrm{nc}}$ can be replaced by a vertex whose vertex group is a triangle group of the form $\Delta(a,a,b)$ $(a,b \in \mathbb{N})$: We know $g(\mathcal{G}^{\mathrm{nt}}) \leq 1$, hence $\chi(\mathcal{G}^{\mathrm{nt}}) \geq 0$, and the same holds for $\mathcal{G}^{\mathrm{nc}}$. Theorem 4.5 states $\chi(\mathcal{G}^{\mathrm{nc}}) + 2\chi(\mathcal{G}^{\mathrm{nt}}) = 1$, therefore $\chi(\mathcal{G}^{\mathrm{nc}}) = 1$ and $\chi(\mathcal{G}^{\mathrm{nt}}) = 0$. As $g(\mathcal{G}^{\mathrm{nt}}) \leq 1$ we conclude $g(\mathcal{G}^{\mathrm{nt}}) = 1$ and $\mathcal{G}^{\mathrm{nt}}$ is connected. Prop. 4.9 states $g(\mathcal{G}^{\mathrm{nc}}) = 0$, therefore $\mathcal{G}^{\mathrm{nc}}$ is connected as well. By Theorem 4.11 we can replace $\mathcal{G}^{\mathrm{nc}}$ by a single vertex v whose vertex group is a triangle group Δ. If we contract the rest of the graph as much as possible, we get an edge from v to v with a cyclic stabilizer. This stabilizer occurs therefore on two ends of $\mathcal{G}^*(\Delta)$ (this was $\mathcal{N}^*$ in Theorem 4.11). □

Note we have seen in Theorem 4.12 that the spherical triangle groups of type $\Delta(a,a,b)$ are $D_n = \Delta(2,2,n)$ and $A_4 = \Delta(2,3,3)$, and for $p > 5$ there exist no other ones. For $p \leq 5$ there are additional possible triangle groups, for $p = 5$ we have seen the type $\Delta(3,3,5n)$ in Example 4.13.

We now know that normal p-adic origamis are always of the form $\Omega/\Gamma \to \Omega/G$, and we know quite well which groups G can occur. It remains to investigate what groups Γ are possible. The only restriction we have for Γ is that it has to be a Schottky group of finite index and normal in G: As the covering $\Omega \to \Omega/\Gamma$ is always unramified the ramification of $\Omega/\Gamma \to \Omega/G$ is equal to the ramification of $\Omega \to \Omega/G$ and hence only depends on G. The genus of Ω/G also does not depend on the choice of Γ.

Theorem 4.15. *Let $G \subset \mathrm{PGL}_2(\mathbb{C}_p)$ be a finitely generated discontinuous group and Γ be a normal subgroup of G of finite index. Then the following statements are equivalent:*

i) Γ is a Schottky group

ii) $\Gamma \cap G_i = \{1\}$ for every vertex group G_i in $\mathcal{G}^*(G)$.

Proof. i) $\Rightarrow$ ii) is easy: The vertex groups are finite, therefore every $g \in G_i$ has finite order. Γ does not contain elements of finite order. Thus $\Gamma \cap G_i$ is trivial. For ii) $\Rightarrow$ i) we proceed with three steps:

Step 1: Every element of Γ has infinite order: G is the fundamental group of $\mathcal{G}^*(G)$, hence an HNN-extension of an amalgamated product of the G_i. If $g \in G$ has finite order $n > 1$, then g is conjugated to a $g' \in G_i$ (see [LS], IV.2.4 and IV.2.7) with $\mathrm{ord}(g') = \mathrm{ord}(g) > 1$. But by assumption $g' \notin \Gamma$ and hence $g \notin \Gamma$ as Γ is normal in G.

Step 2: Γ is free by Ihara's theorem ([Ser], I.1.5, Theorem 4): G acts on the tree $\mathcal{T}^*(G)$ with quotient graph $\mathcal{G}^*(G)$. For an $x \in \mathcal{T}^*(G)$ let $g \in \mathrm{Stab}_G(x)$ be non-trivial. Then g has finite order, and with step 1 we see $g \notin \Gamma$. Therefore the action on $\mathcal{T}^*(G)$ restricted to Γ is free, thus Γ is a free group by [Ser], §3.3.

Step 3: Γ is a Schottky group: Γ is by definition a finite index normal subgroup of G. It is discontinuous because G is and it contains no elements of finite order. It is finitely generated because G is (by Reidemeister-Schreier, [LS], Prop. II.4.2). Thus we know that Γ is a Schottky group. We can even find a finite set of free generators of Γ by looking at its action on $\mathcal{T}^*(G)$: this action has a finite fundamental domain and this domain therefore has only finitely many neighboring translates. The set of these neighboring translates corresponds to a finite set of free generators for Γ. □

We are especially interested in the resulting Galois group $H := G/\Gamma$. Thus we now answer the question what choices of Γ are possible if we fix this Galois group:

Corollary 4.16. *Let H be a finite group and $G \subset \mathrm{PGL}_2(\mathbb{C}_p)$ be a finitely generated discontinuous group. Further let Γ be the kernel of a homomorphism $\varphi : G \to H$. Then the following statements are equivalent:*

i) Γ *is a Schottky group*

ii) $\varphi|_{G_i}$ *is injective for every vertex group* G_i *in* $\mathcal{G}^*(G)$.

Proof. Follows with $\ker(\varphi|_{G_i}) = \Gamma \cap G_i$ from the Theorem. □

Example 4.17. a) Set $G_n := \langle D_n, \gamma;\, \gamma\sigma = \sigma\gamma\rangle$ as in Example 4.3 (with n odd and $\mathrm{ord}(\sigma) = 2$). We extend this example: Choose $m \in \mathbb{N}$ and define $\varphi : G_n \to D_n \times C_m$ by $\varphi|_{D_n} = (\mathrm{id}, 1)$ and $\varphi(\gamma) = (1, c)$ where c is a generator of C_m. Then Corollary 4.16 states that $\Gamma' := \ker(\varphi)$ is a Schottky group and the Galois group of the p-adic origami $\Omega/\Gamma' \to \Omega/G$ is $D_n \times C_m$. Note that $\Gamma' \subseteq \Gamma := \ker(G_n \to D_n)$ for every m, thus we have a covering of origamis $\Omega/\Gamma' \to \Omega/\Gamma \to \Omega/G$. We will investigate such coverings in Section 4.5.

b) Set $G := \langle A_4, \gamma;\, \gamma\delta = \delta\gamma\rangle$ as in Example 4.4 (with $\mathrm{ord}(\delta) = 3$). We extend this example as in a): Choose $m \in \mathbb{N}$ and define $\varphi : G_n \to A_4 \times C_m$ by $\varphi|_{A_4} = (\mathrm{id}, 1)$ and $\varphi(\gamma) = (1, c)$ where c is a generator of C_m. Then Corollary 4.16 states that $\Gamma' := \ker(\varphi)$ is a Schottky group and the Galois group of the p-adic origami $\Omega/\Gamma' \to \Omega/G$ is $A_4 \times C_m$. Note that $\Gamma' \subseteq \Gamma := \ker(G_n \to A_4)$ for every m, thus we have again a covering of origamis $\Omega/\Gamma' \to \Omega/\Gamma \to \Omega/G$.

c) In Example 4.13 we have constructed the 5-adic triangle group $\Delta(3, 3, 5) = A_5 *_{D_5} A_5$. The group $G := \langle A_5 *_{D_5} A_5, \gamma;\, \gamma\delta_1 = \delta_2\gamma\rangle$, where the δ_i of order 3 are chosen out of the two different A_5-components, can be embedded into $\mathrm{PGL}_2(\mathbb{Q}_5)$ (to show this one can use [Kat2], Theorem II). Then we can define $\varphi : G \to A_5$ as the identity on both A_5-components of the amalgamated product, and $\varphi(\gamma) = 1$. This leads to a 5-adic origami with Galois group A_5.

d) Take the group G_5 from part a) and consider the homomorphism $\varphi : G_5 \to \mathrm{PSL}_2(\mathbb{F}_{11})$ defined by

$$\varphi(\sigma) = \begin{pmatrix} 6 & 7 \\ 1 & 5 \end{pmatrix}, \quad \varphi(\delta) = \begin{pmatrix} 0 & 9 \\ 6 & 3 \end{pmatrix}, \quad \varphi(\gamma) = \begin{pmatrix} 3 & 7 \\ 8 & 8 \end{pmatrix}$$

We can calculate that this is indeed a homomorphism by checking the relations of G_5 for the given images of φ. We can also check that φ is surjective and that $\varphi|_{D_5}$ is injective. Hence $\ker(\varphi)$ is a Schottky group and the corresponding origami has Galois group $\mathrm{PSL}_2(\mathbb{F}_{11})$.

4.4 Automorphisms of p-adic origamis

The Galois group of the covering $X \to E$ is a subgroup of the automorphism group $\operatorname{Aut} X$. For normal complex origamis we know from Proposition 3.12 that the Galois group consists precisely of all possible translations. In Proposition 3.13 we have seen that if the automorphism group is strictly larger than the Galois group, then there have to be automorphisms which are not translations, i.e. there is an automorphism which does not induce the identity but an involution on E. We now investigate the implications if this happens in the p-adic setting.

Theorem 4.18. *In the situation of Theorem 4.14 let* $\operatorname{Aut}(X)$ *contain an element* σ *of order 2 which induces a non-trivial automorphism* $\bar{\sigma}$ *of* E *fixing the branch point of* $X \to E$*. Then there is a discontinuous group* H *containing* G *as normal subgroup of index 2, which is isomorphic to the fundamental group of the graph of groups*

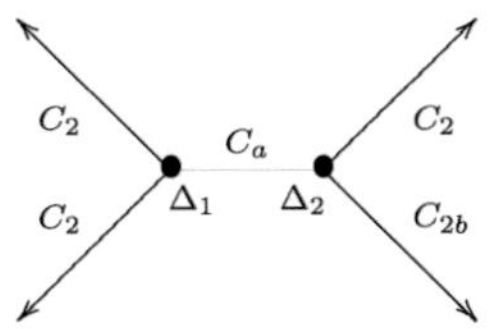

where Δ_1 *is the p-adic triangle group of type* $\Delta(2,2,a)$*, i.e.* $\Delta_1 \cong D_a$*, and* Δ_2 *is a p-adic triangle group of type* $\Delta(2,a,2b)$ *containing* Δ *of index 2.*

Proof. Let L be the subgroup of $\operatorname{Aut} X$ generated by σ and the Galois group $\operatorname{Gal}(X/E)$. Every $\ell \in L \setminus \operatorname{Gal}(X/E)$ induces $\bar{\sigma}$ on E, thus $\ell \circ \sigma \in \operatorname{Gal}(X/E)$. Hence L contains $\operatorname{Gal}(X/E)$ with index 2, and therefore as a normal subgroup. As in the proof of Theorem 4.14 we have $L \cong H/\Gamma$ and for a discontinuous group H and $\operatorname{Gal}(X/E) \cong G/\Gamma$ for normal subgroup G of index 2 in H with $\Omega(H) = \Omega(G) = \Omega(\Gamma)$. Now $\Omega/H \cong E/\langle\bar{\sigma}\rangle =: P$.

The branch point of $X \to E$ is a fixed point of $\bar{\sigma}$ and therefore a ramification point of $E \to P$. By Riemann-Hurwitz this means that $g(P) = 0$ and there are four ramification points of $E \to P$. As the degree of the map $E \to P$ is 2, and this is also the ramification index of the four ramification points, we know that there are also exactly four branch points. The composition $X \to P$ of the maps thus has four branch points. Over three of them the map $X \to E$ is unramified, therefore the corresponding ramification indices of $X \to P$ are 2. Over the fourth branch point the map $X \to E$ is ramified with ramification index b, thus the total ramification index is $2b$.

Now let $\mathcal{H}^*$ be the quotient graph corresponding to H. Since $\Omega/H \cong P$ the graph $\mathcal{H}^*$ is a realizable graph of genus zero with four ends. The stabilizer of one end is a cyclic group of order $2b$, the stabilizers of the other three ends are cyclic groups of order 2.

We have $g(\mathcal{H}^*) = 0$, thus this also holds for all subgraphs of $\mathcal{H}^*$. Therefore any Euler characteristic equals the number of connected components and we have $\chi(\mathcal{H}^{\text{nc}}) \geq 1$ and $\chi(\mathcal{H}^{\text{nt}}) \geq 1$. As $\chi(\mathcal{H}^{\text{nc}}) + 2\chi(\mathcal{H}^{\text{nt}}) = 4$ by Theorem 4.5, we conclude $\chi(\mathcal{H}^{\text{nc}}) = 2$ and $\chi(\mathcal{H}^{\text{nt}}) = 1$. Thus $\mathcal{H}^{\text{nc}}$ has two connected components, which we can by Theorem 4.11 both replace by single vertices whose vertex groups are triangle groups Δ_1 and Δ_2. Furthermore $\mathcal{H}^{\text{nt}}$ has to be connected, therefore those two triangle groups have to be connected by a path with a nontrivial cyclic stabilizer. This path can be contracted to a single edge.

It remains to find the connection between the stabilizing groups of $\mathcal{G}^*$ and $\mathcal{H}^*$. We get the graph $\mathcal{H}^*$ as the quotient of $\mathcal{G}^*$ by σ (if we ignore the ends of both graphs). The single edge from $\mathcal{G}^*$ has to be mapped to itself and inverted (because otherwise there would still be a closed edge in $\mathcal{H}^*$). Thus we have to insert a vertex on this edge and for constructing the quotient $\mathcal{G}^*$ we have to take one of the half edges. Therefore the stabilizer of this edge (as it is not fixed by σ) is the same as before (namely C_a) and the original vertex is fixed by Δ and σ. □

Example 4.19. Let ζ be a primitive 10-th root of unity and choose δ, σ and γ as in Example 4.3. Then the group $G := \langle \delta^2, \sigma, \gamma \rangle$ corresponds to the case $n = 5$ of this example. But if we work out the quotient graph of $H := \langle \delta, \sigma, \gamma \rangle$ we note that while there still is a vertex with stabilizer $\langle \sigma, \delta \rangle \cong D_{10}$ now the element $\gamma\delta^5$ is elliptic of order 2, but does not fix this vertex. Instead there is now another vertex fixed by $\gamma\delta^5$ on the axis of σ, its stabilizer is therefore $\langle \sigma, \gamma\delta^5 \rangle \cong D_2$. This means that γ does not create a circle in the graph any more, but the quotient graph becomes

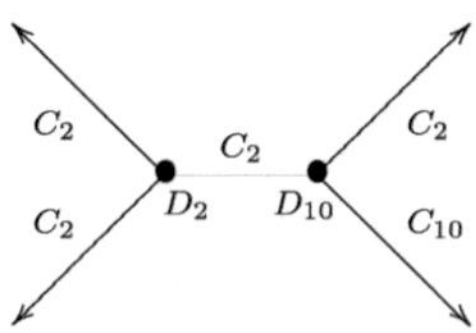

Now we define a homomorphism $\varphi : H \to \mathrm{PSL}_2(\mathbb{F}_{11}) \times \mathbb{Z}/2\mathbb{Z}$ by

$$\varphi(\sigma) = \left(\begin{pmatrix} 6 & 7 \\ 1 & 5 \end{pmatrix}, 0 \right), \quad \varphi(\delta) = \left(\begin{pmatrix} 8 & 6 \\ 4 & 10 \end{pmatrix}, 1 \right), \quad \varphi(\gamma) = \left(\begin{pmatrix} 3 & 7 \\ 8 & 8 \end{pmatrix}, 0 \right).$$

As in Example 4.17 d) we can check that this is indeed a homomorphism and is injective when restricted to the vertex groups. Note that $\varphi(\delta^2) = (\left(\begin{smallmatrix} 0 & 9 \\ 6 & 3 \end{smallmatrix}\right), 0)$, thus $\varphi|_G$ is exactly the homomorphism considered in Example 4.17 d), and we have $\ker(\varphi) = \ker(\varphi|_G) = \Gamma$. Thus the p-adic origami $\Omega/\Gamma \to \Omega/G$ can be extended to $\Omega/\Gamma \to \Omega/H$ with Galois group $H/\Gamma \cong \mathrm{PSL}_2(\mathbb{F}_{11}) \times \mathbb{Z}/2\mathbb{Z}$. This means that the automorphism group of this origami contains the group $\mathrm{PSL}_2(\mathbb{F}_{11}) \times \mathbb{Z}/2\mathbb{Z}$.

4.5 Coverings of p-adic origamis

Definition 4.20. We call a p-adic origami $\Omega/\Gamma' \to \Omega/G$ *simple* if it covers one of the origamis from Examples 4.3 and 4.4. This means we have $\Omega/\Gamma' \to \Omega/\Gamma \to \Omega/G$ where

$$G = \langle \Delta = D_n, \gamma; \gamma\sigma = \sigma\gamma \rangle \text{ with } n \in \mathbb{N} \text{ odd and } \sigma \in D_n \text{ of order 2 or}$$
$$G = \langle \Delta = A_4, \gamma; \gamma\delta = \delta\gamma \rangle \text{ with } \delta \in A_4 \text{ of order 3.}$$

Γ and Γ' are free normal subgroups of finite index in G, and $G/\Gamma \cong \Delta$.

Theorem 4.21. *Let G, Δ and γ be given as in the definition above, $\varphi : \Delta \to \mathrm{Aut}(K)$ an action of Δ on a finite group K and $\overline{\gamma} \in K$ an element which is fixed by all $\delta \in \Delta$ which commute with γ in G.*

Define $\psi : G \to \Delta \ltimes_\varphi K$ by $\delta \mapsto (\delta, 1)$ for all $\delta \in \Delta$ and $\gamma \mapsto (1, \overline{\gamma})$ and set $\Gamma' = \ker(\psi)$. Then $\Omega/\Gamma' \to \Omega/G$ is a simple p-adic origami with Galois group $\mathrm{Im}(\psi)$.

Every simple p-adic origami is of this type.

Proof. "$\Rightarrow$": Let K, φ and $\overline{\gamma}$ be given. We have homomorphisms

$$G \xrightarrow{\psi} \Delta \ltimes_\varphi K \xrightarrow{p} \Delta$$

where p is the projection to the first component. As $\psi|_\Delta$ is injective Γ' is a Schottky group by Corollary 4.16. We see $\Gamma = \ker(p \circ \psi)$, thus $\Gamma' \subset \Gamma$, therefore we have homomorphisms $\Omega/\Gamma' \to \Omega/\Gamma \to \Omega/G$. The Galois group of the composed covering is $G/\Gamma' = G/\ker(\psi) \cong \mathrm{Im}(\psi)$.

"$\Leftarrow$": Let $\Omega/\Gamma' \to \Omega/G$ be a simple p-adic origami. We have homomorphisms

$$\langle \Delta, \gamma \rangle = G \xrightarrow{\alpha} G/\Gamma' \xrightarrow{\beta} (G/\Gamma')\Big/(\Gamma/\Gamma') \cong G/\Gamma \cong \Delta$$

where $\beta \circ \alpha$ is an isomorphism on Δ and maps γ to 1. Therefore $\beta(\alpha(\gamma)) = 1$, thus $\alpha(\gamma) \in \ker(\beta) =: K$. Moreover $\beta|_{\alpha(\Delta)}$ is an isomorphism. We get a split exact sequence

$$0 \longrightarrow K \longrightarrow G/\Gamma' \longrightarrow \Delta \longrightarrow 0$$

Thus there is an isomorphism from G/Γ' to a semidirect product $\Delta \ltimes_\varphi K$ where $\varphi : \Delta \to \mathrm{Aut}(K)$ is the homomorphism induced by the split exact sequence, i.e. conjugation by $\alpha(\Delta)$. The restriction of this isomorphism to Δ is given by the inclusion $\delta \mapsto (\delta, 1)$. The image of γ is in the kernel of p, hence of the form $(1, \overline{\gamma})$ for a $\overline{\gamma} \in K$.

Now let $\delta \in \Delta$ be given with $\gamma\delta = \delta\gamma$. In $\Delta \ltimes_\varphi K$ this means

$$(\delta, \overline{\gamma}) = (1, \overline{\gamma})(\delta, 1) = (\delta, 1)(1, \overline{\gamma}) = (\delta, \varphi(\delta)(\overline{\gamma}))$$

Therefore $\varphi(\delta)$ fixes $\overline{\gamma}$. □

Corollary 4.22. *Let $\Omega/\Gamma' \to \Omega/\Gamma$ be a simple p-adic origami with $\delta\gamma = \gamma\delta$ in G/Γ' for all $\delta \in \Delta$. Then the action in the last theorem is trivial on $\overline{\gamma}$, thus the Galois group of this origami is $\Delta \times \mathbb{Z}/m\mathbb{Z}$ for an $m \in \mathbb{N}$.*

Example 4.23. Set $G = \langle \Delta = D_n, \gamma; \gamma\sigma = \sigma\gamma \rangle$ with $n \in \mathbb{N}$ odd and $\sigma \in D_n$ of order 2. Δ acts on the set $\{1, \ldots, n\}$. As n is odd σ has a fixed point k.

Now choose $m \in \mathbb{N}$, set $K := (\mathbb{Z}/m\mathbb{Z})^n$, let $\overline{\gamma} := e_k \in K$ be the k-th unit vector and define $\varphi : \Delta \to K$ by $\delta \mapsto ((x_i) \mapsto (x_{\delta(i)}))$. By the theorem this defines a simple p-adic origami $\Omega/\Gamma' \to \Omega/G$.

As $G \to \Delta \ltimes_\varphi K$ is surjective, the Galois group of this origami is $\Delta \ltimes_\varphi K$.

The same works for $G = \langle \Delta = A_4, \gamma; \gamma\delta = \delta\gamma \rangle$ with $\delta \in A_4$ of order 3: Δ acts on $\{1, 2, 3, 4\}$ and δ has a fixed point. Thus we can define a simple p-adic origami with Galois group $\Delta \ltimes_\varphi (\mathbb{Z}/m\mathbb{Z})^4$.

Chapter 5

Connecting both worlds

In the first three Chapters we have investigated origamis over $\mathbb{C}$, while in Chapter 4 we defined origamis also over $\mathbb{C}_p$. Now we want to connect both worlds: An origami-curve in $\mathcal{M}_{g,\mathbb{C}}$ is always defined over $\overline{\mathbb{Q}}$, and thus defines also a curve in $\mathcal{M}_{g,\overline{\mathbb{Q}}}$, and this curve can in turn be interpreted as a curve in $\mathcal{M}_{g,\mathbb{C}_p}$. Now we ask the question: Does this curve intersect the subspace of $\mathcal{M}_{g,\mathbb{C}_p}$ containing Mumford curves?
The resulting points in $\mathcal{M}_{g,\mathbb{C}_p}$ are still curves which cover an elliptic curve with only one branch point. Thus those curves are Mumford curves if and only if they occur as p-adic origamis. We have introduced several invariants of origami-curves in the last chapters, some of which turn up in both worlds: The ramification indices, the Galois group of a normal origami and its automorphism group. By the Lefschetz principle ([Lef], Appendix) these algebraic properties of the origamis coincide over $\mathbb{C}$ and over $\mathbb{C}_p$. In some cases this is enough information to identify the complex origami-curve which belongs to a given p-adic origami.

5.1 Base change of schemes

Let $X_{\mathbb{C}} \to E_{\mathbb{C}}$ be an origami over $\mathbb{C}$. We can write $E_{\mathbb{C}} = E_{\mathbb{C},\tau} = \mathbb{C}/(\mathbb{Z}+\tau\mathbb{Z})$ with $\tau \in \mathbb{H}$, where 0 is the only branch point of $X_{\mathbb{C}} \to E_{\mathbb{C}}$. We have the Weierstrass-covering $\wp : E_{\mathbb{C},\tau} \to \mathbb{P}^1(\mathbb{C})$. For any ramification point $x \neq 0$ we have $\wp'(x)^2 = 4\wp^3(x) - g_2\wp(x) - g_3 = 0$, so if $g_2, g_3 \in \overline{\mathbb{Q}}$ we get $\wp(x) \in \overline{\mathbb{Q}}$. As $\wp(0) = \infty$ Belyi's theorem would then imply that both $X_{\mathbb{C}}$ and $E_{\mathbb{C}}$ are defined over $\overline{\mathbb{Q}}$. Therefore we then have $X_{\overline{\mathbb{Q}}}$ and $E_{\overline{\mathbb{Q}}}$ over $\overline{\mathbb{Q}}$ with the following diagram of base changes:

$$\begin{array}{ccccc} X_{\overline{\mathbb{Q}}} \times_{\overline{\mathbb{Q}}} \mathbb{C} = X_{\mathbb{C}} & \longrightarrow & X_{\overline{\mathbb{Q}}} & \longleftarrow & X_{\mathbb{C}_p} = X_{\overline{\mathbb{Q}}} \times_{\overline{\mathbb{Q}}} \mathbb{C}_p \\ \downarrow & & \downarrow & & \downarrow \\ E_{\overline{\mathbb{Q}}} \times_{\overline{\mathbb{Q}}} \mathbb{C} = E_{\mathbb{C}} & \longrightarrow & E_{\overline{\mathbb{Q}}} & \longleftarrow & E_{\mathbb{C}_p} = E_{\overline{\mathbb{Q}}} \times_{\overline{\mathbb{Q}}} \mathbb{C}_p \end{array}$$

By varying $\tau \in \mathbb{H}$ we get a curve in the moduli-space $\mathcal{M}_{g,\mathbb{C}}$, which leads to a curve in $\mathcal{M}_{g,\overline{\mathbb{Q}}}$, which itself can be considered as a subset of a curve in $\mathcal{M}_{g,\mathbb{C}_p}$. This curve may or may not intersect the subset of Mumford curves in $\mathcal{M}_{g,\mathbb{C}_p}$.

Example 5.1. Consider the origami from Example 3.19:

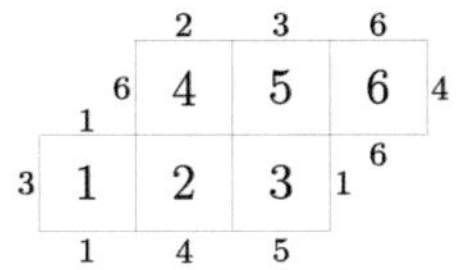

Kappes has proven in [Kap], Theorem IV.3.7 that the origami-curve in $\mathcal{M}_{g,\mathbb{C}}$ of this origami contains all the curves birationally equivalent to

$$y^2 = (x^2-1)(x^2-\lambda^2)\left(x^2 - \left(\tfrac{\lambda}{\lambda+1}\right)^2\right)$$

for $\lambda \in \mathbb{C} \setminus \left\{0, \pm 1, -\frac{1}{2}, -2\right\}$. If we now restrict the choice of λ to $\overline{\mathbb{Q}}$, we get a curve in $\mathcal{M}_{g,\overline{\mathbb{Q}}}$. We can now change the base of this curve to $\mathbb{C}_p$ for an arbitrary prime p. This will result in a curve in $\mathcal{M}_{g,\mathbb{C}_p}$. Does this curve intersect the subspace of $\mathcal{M}_{g,\mathbb{C}_p}$ containing Mumford curves?

Fortunately [Bra3], Theorem 4.3 offers a criterion for a hyperelliptic curve X to be a Mumford curve: This is the case if and only if the branch points of $X \to \mathbb{P}^1$ (in our case $\pm 1, \pm\lambda$ and $\pm\frac{\lambda}{\lambda+1}$) can be matched into pairs (a_i, b_i) such that $\mathbb{P}^1$ can be covered by annuli U_i each containing exactly one of those pairs. In our case we consider only $p > 2$, set $\lambda := q-1$ for any $q \in \mathbb{C}_p^\times$ with $|q| < 1$ and match the points as follows:

$$\begin{array}{lll} a_1 = 1, & b_1 = -\lambda = 1-q & \Rightarrow |a_1 - b_1| = |q| < 1 \\ a_2 = -1, & b_2 = \lambda = q-1 & \Rightarrow |a_2 - b_2| = |q| < 1 \\ a_3 = \frac{\lambda}{\lambda+1} = \frac{1}{q}(q-1), & b_3 = -\frac{1}{q}(q-1) & \Rightarrow |a_3| = |b_3| = \left|\frac{1}{q}\right| > 1 \end{array}$$

Thus we can choose $U_1 = B(1,1) \setminus B(-1,|q|)$, $U_2 = B(1,1) \setminus B(1,|q|)$ and $U_3 = \mathbb{P}^1 \setminus B(1,1)$ to get the desired covering[1].

[1]The balls $B(x,r)$ were defined in Section 4.1.

In general it is almost impossible to find the equation of a given origami. Therefore we would like a simpler approach, and we will do this the other way round: We have already constructed p-adic origamis, now we try to match them to the corresponding complex origami-curve. We will do this by matching some of the invariants we have calculated for both of them in the previous chapters.

5.2 Galois groups with a unique curve

Some Galois groups occur only for a single origami-curve over $\mathbb{C}$. We will now prove that this is the case for the Galois groups $D_n \times \mathbb{Z}/m\mathbb{Z}$ and $A_4 \times \mathbb{Z}/m\mathbb{Z}$, which occurred as Galois groups of p-adic origamis in Example 4.17 a) and b).

Lemma 5.2. *Let $f : F_2 \to \mathbb{Z}/m\mathbb{Z}$ be surjective. Then up to an automorphism of F_2 we can assume $f(x) = 1$ and $f(y) = 0$.*

Proof. Let $c_x, c_y \in \mathbb{N}$ with $c_x = f(x)$ and $c_y = f(y)$ in $\mathbb{Z}/m\mathbb{Z}$.
We prove first that we can choose the representatives c_x and c_y coprime: Let p_i be the prime factors of c_x and set

$$c'_y := c_y + m \cdot \prod_{p_i \nmid c_y} p_i$$

Assume that there is a p_i which is a factor of c'_y. If $p_i \mid c_y$ then p_i would also have to be a factor of the right-hand summand, and as it is not contained in the product we would then have $p_i \mid m$. But this contradicts $\langle c_x, c_y \rangle = \mathbb{Z}/m\mathbb{Z}$. If on the other hand $p_i \nmid c_y$, then p_i would be a factor of the right-hand summand but not of the left one, which would contradict the assumption.
Therefore no p_i is a factor of c'_y, thus $\gcd(c_x, c'_y) = 1$. We can replace c_y by c'_y as both are equivalent modulo m.
Now $\gcd(c_x, c_y) = 1$, thus there exist $a, b \in \mathbb{Z}$ with $ac_x + bc_y = 1$. Set

$$A := \begin{pmatrix} a & -c_y \\ b & c_x \end{pmatrix} \in \mathrm{SL}_2(\mathbb{Z})$$

and let $\varphi \in \mathrm{Aut}(F_2)$ be a lift of A. As $\mathbb{Z}/m\mathbb{Z}$ is abelian we get a commutative diagram

$$\begin{array}{ccccc} F_2 & \xrightarrow{\varphi} & F_2 & \xrightarrow{f} & \mathbb{Z}/m\mathbb{Z} \\ \searrow & & \searrow & \nearrow_{\bar{f}} & \\ \mathbb{Z}^2 & \xrightarrow{\bar{\varphi}} & \mathbb{Z}^2 & & \end{array}$$

where $\overline{\varphi}$ is the multiplication with A. Thus

$$(f \circ \varphi)(x) = (\overline{f} \circ \overline{\varphi})\left(\begin{pmatrix}1\\0\end{pmatrix}\right) = \overline{f}\left(\begin{pmatrix}a\\b\end{pmatrix}\right) = ac_x + bc_y = 1$$
$$(f \circ \varphi)(y) = (\overline{f} \circ \overline{\varphi})\left(\begin{pmatrix}0\\1\end{pmatrix}\right) = \overline{f}\left(\begin{pmatrix}-c_y\\c_x\end{pmatrix}\right) = -c_yc_x + c_xc_y = 0$$

□

Proposition 5.3. *Let $n, m \in \mathbb{N}$ and n be odd. Up to an automorphism of F_2 there exists only one origami with Galois group $D_n \times \mathbb{Z}/m\mathbb{Z}$.*

Proof. Choose $\sigma, \tau \in D_n$ with $\langle\sigma, \tau\rangle = D_n$ and $\mathrm{ord}(\sigma) = \mathrm{ord}(\tau) = 2$, and set $\delta = \sigma\tau$.

Let $f : F_2 \to D_n \times \mathbb{Z}/m\mathbb{Z}$ be the monodromy of a normal origami. By Lemma 5.2 we can apply an automorphism of F_2 to get $f(x) = (\alpha, 1)$ and $f(y) = (\beta, 0)$ for some $\alpha, \beta \in D_n$. As f has to be surjective we know $\langle\alpha, \beta\rangle = D_n$. Up to an automorphism of D_n we have three cases:

i) $\mathrm{ord}(\alpha) = 2, \mathrm{ord}(\beta) = 2$, w.l.o.g. $(\alpha, \beta) = (\tau, \sigma)$

ii) $\mathrm{ord}(\alpha) = n, \mathrm{ord}(\beta) = 2$, w.l.o.g. $(\alpha, \beta) = (\delta, \sigma)$

iii) $\mathrm{ord}(\alpha) = 2, \mathrm{ord}(\beta) = n$, w.l.o.g. $(\alpha, \beta) = (\tau, \delta)$

We can apply $\varphi \in \mathrm{Aut}(F_2)$ with $x \mapsto yx$ and $y \mapsto y$ to get from i) to ii):

$$(f \circ \varphi)(x) = f(yx) = (\sigma\tau, 0 + 1) = (\delta, 1) \text{ and}$$
$$(f \circ \varphi)(y) = f(y) = (\sigma, 0)$$

For odd m we can apply $\varphi \in \mathrm{Aut}(F_2)$ with $x \mapsto x$ and $y \mapsto yx^m$ to get from i) to iii):

$$(f \circ \varphi)(x) = f(x) = (\tau, 1) \text{ and}$$
$$(f \circ \varphi)(y) = f(yx^m) = (\sigma\tau^m, 0 + m) = (\delta, 0)$$

For even m case iii) is not possible, as f would not be surjective: Assume there is a preimage $z \in f^{-1}(\mathrm{id}, 1)$. We know that $f(xy) = f(y^{-1}x)$ (because this holds in both components). Thus we can choose z of the form $x^a y^b$, its image is $(\tau^a\delta^b, a)$. Now for $f(z)_1 = \mathrm{id}$ we need a even, but for $f(z)_2 = 1$ we need $a = 1$ in $\mathbb{Z}/m\mathbb{Z}$. For even m this is impossible. Therefore there is no preimage of $(\mathrm{id}, 1)$ in this case. □

Proposition 5.4. *Up to an automorphism of F_2 there exists, for any $m \in \mathbb{N}$, only one origami with Galois group $A_4 \times \mathbb{Z}/m\mathbb{Z}$.*

Proof. Let $f : F_2 \to A_4 \times \mathbb{Z}/m\mathbb{Z}$ be the monodromy of a normal origami. By Lemma 5.2 we can apply an automorphism of F_2 to get $f(x) = (\alpha, 1)$ and $f(y) = (\beta, 0)$. As f has to be surjective we know $\langle \alpha, \beta \rangle = A_4$. Up to an isomorphism of A_4 we have four cases:

i) $\mathrm{ord}(\alpha) = 2, \mathrm{ord}(\beta) = 3$, w.l.o.g. $(\alpha, \beta) = ((1\,2)(3\,4), (2\,3\,4))$

ii) $\mathrm{ord}(\alpha) = 3, \mathrm{ord}(\beta) = 2$, w.l.o.g. $(\alpha, \beta) = ((1\,2\,3), (1\,2)(3\,4))$

iii) $\mathrm{ord}(\alpha) = \mathrm{ord}(\beta) = 3$, $\mathrm{ord}(\alpha\beta) = 2$, w.l.o.g. $(\alpha, \beta) = ((1\,2\,3), (2\,3\,4))$

iv) $\mathrm{ord}(\alpha) = \mathrm{ord}(\beta) = 3$, $\mathrm{ord}(\alpha\beta) = 3$, w.l.o.g. $(\alpha, \beta) = ((1\,2\,4), (2\,3\,4))$

We can apply $\varphi \in \mathrm{Aut}(F_2)$ with $x \mapsto xy$ and $y \mapsto y$ to get from iii) to i):

$$(f \circ \varphi)(x) = f(xy) = ((1\,2\,3)(2\,3\,4), 1) = ((1\,2)(3\,4), 1) \text{ and}$$
$$(f \circ \varphi)(y) = f(y) = ((2\,3\,4), 0)$$

We can apply $\varphi \in \mathrm{Aut}(F_2)$ with $x \mapsto xy^{-1}$ and $y \mapsto y$ to get from iii) to iv):

$$(f \circ \varphi)(x) = f(xy^{-1}) = ((1\,2\,3)(2\,4\,3), 1) = ((1\,2\,4), 1) \text{ and}$$
$$(f \circ \varphi)(y) = f(y) = ((2\,3\,4), 0)$$

If m is not divisible by 3 choose $k \in \mathbb{N}$ such that $km \equiv 1 \mod 3$. We can then apply $\varphi \in \mathrm{Aut}(F_2)$ with $x \mapsto x$ and $y \mapsto x^{km}y$ to get from iii) to ii):

$$(f \circ \varphi)(x) = f(x) = ((1\,2\,3), 1) \text{ and}$$
$$(f \circ \varphi)(y) = f(x^{km}y) = ((1\,2\,3)(2\,3\,4), 0 + km) = ((1\,2)(3\,4), 0)$$

If m is divisible by 3, case ii) is not possible, as f would not be surjective: Assume there is a preimage $z \in f^{-1}(\mathrm{id}, 1)$. As the first component of $f(z)$ is $\mathrm{id} \in A_4$ the presentation of A_4 tells us that the element z is contained in the normal subgroup generated by y^2, x^3 and $(yx)^3$, and hence is a product of conjugates of those elements. But $f(y^2) = (\mathrm{id}, 0)$ and $f(x^3) = f((yx)^3) = (\mathrm{id}, 3)$. Therefore the second component of $f(z)$ is divisible by 3. As m is also divisible by 3 this contradicts $f(z) = (\mathrm{id}, 1)$. □

Example 5.5. In Example 5.1 we have shown that the origami O defined by

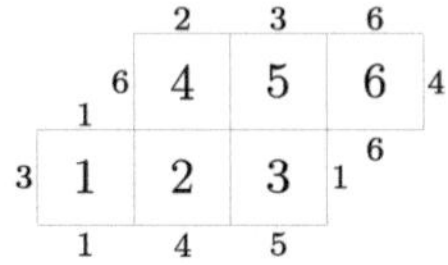

occurs as a p-adic origami. Now we have an alternative way of showing this: Let O' be the minimal normal cover of O studied in Example 3.19. We have seen there that the Galois group of O' is A_4. Proposition 5.4 tells us that the origami-curve of O' is the only curve with this Galois group.

In Example 4.4 we constructed a p-adic origami with Galois group A_4; in fact one for every suitably chosen $q \in \mathbb{P}^1(\mathbb{C}_p)$. If one of those origamis X' is (as an algebraic curve) defined over $\overline{\mathbb{Q}}$ and hence over $\mathbb{C}$ then it has to occur on the origami-curve of O'. The covering $O' \to O$ leads to a morphism $X' \to X$, where X is on the origami-curve of O. As X' is a Mumford curve the same holds for X by [Bra1], Satz 5.24.

5.3 When the group is not enough

In section 2.2 we have proposed an algorithm to find all complex origami-curves of normal origamis with a given Galois group H. This results in a set of representative origamis given as epimorphisms $f : F_2 \twoheadrightarrow H$ as described in Proposition 1.6. For most groups there is only one curve with the given Galois group, but there are cases where there are more than one[2]. The smallest example for such a group is the group A_5 where there are two curves. Representatives are given by

$$\begin{array}{llll} f_1 : F_2 \to A_5, & x \mapsto (1\,5\,3\,4\,2), & y \mapsto (1\,3\,2\,4\,5) & \text{and} \\ f_2 : F_2 \to A_5, & x \mapsto (1\,5\,2\,4\,3), & y \mapsto (2\,3)(4\,5) & \end{array}$$

We can easily see that those two origamis do not define the same curve: In Corollary 3.3 we have seen that the ramification index of a normal origami is the order of $f(xyx^{-1}y^{-1})$. Hence the ramification indices of those two origamis are

$$\operatorname{ord}(f_1(xyx^{-1}y^{-1})) = 5 \text{ and } \operatorname{ord}(f_2(xyx^{-1}y^{-1})) = 3$$

[2]There are 2386 groups with order less than or equal to 250 which can be generated by two elements. Of these there are only 30 where there is more than one curve with this Galois group. We list those groups in Appendix A.3.

Example 5.6. In Example 4.17 c) we have investigated a 5-adic origami with Galois group A_5. Our 5-adic origami had ramification index 5 (recall that this was the order of the cyclic stabilizer of the single end of the quotient graph), hence this corresponds to the curve of the origami defined by f_1.

Sometimes fixing the ramification index makes the origami-curve unique. But there are still some cases where two curves have equal Galois groups and equal ramification index. An example for such a group is the group $\mathrm{PSL}_2(\mathbb{F}_7)$, where there are even four curves, represented by $f_i : F_2 \to \mathrm{PSL}_2(\mathbb{F}_7)$ with $f_i(x) = \sigma_i$ and $f_i(y) = \tau_i$ with

$$\sigma_1 = \sigma_2 = \sigma_3 = \begin{pmatrix} 1 & 2 \\ 5 & 4 \end{pmatrix}, \quad \sigma_4 = \begin{pmatrix} 0 & 6 \\ 1 & 6 \end{pmatrix}$$
$$\tau_1 = \begin{pmatrix} 6 & 0 \\ 2 & 6 \end{pmatrix}, \quad \tau_2 = \sigma_4, \quad \tau_3 = \begin{pmatrix} 1 & 3 \\ 6 & 5 \end{pmatrix}, \quad \tau_4 = \begin{pmatrix} 6 & 3 \\ 3 & 4 \end{pmatrix}$$

The two curves defined by f_1 and f_2 both have ramification index 4.

We have seen in Proposition 3.13 that the automorphism group of an origami is isomorphic to $C_2 \ltimes_\Phi H$ where $\Phi : C_2 \to \mathrm{Aut}(H)$ maps the generator of the cyclic group C_2 to the automorphism φ of H defined by $\varphi(f(x)) = f(x)^{-1}$ and $\varphi(f(y)) = f(y)^{-1}$. In the example above with Galois group $\mathrm{PSL}_2(\mathbb{F}_7)$ and ramification index 4 the automorphism groups are isomorphic and hence can not be used to distinguish those two curves. But in some cases they are helpful:

Example 5.7. In Example 4.17 d) we have investigated a 5-adic origami with Galois group $\mathrm{PSL}_2(\mathbb{F}_{11})$ with ramification index 5. There are four possible origami-curves, represented by $f_i : F_2 \to \mathrm{PSL}_2(\mathbb{F}_{11})$ with $f_i(x) = \sigma_i$ and $f_i(y) = \tau_i$ with

$$\sigma_1 = \sigma_2 = \sigma_3 = \begin{pmatrix} 4 & 6 \\ 8 & 4 \end{pmatrix}, \quad \sigma_4 = \begin{pmatrix} 10 & 10 \\ 7 & 6 \end{pmatrix}$$
$$\tau_1 = \sigma_4, \quad \tau_2 = \begin{pmatrix} 2 & 1 \\ 8 & 10 \end{pmatrix}, \quad \tau_3 = \begin{pmatrix} 1 & 4 \\ 0 & 1 \end{pmatrix}, \quad \tau_4 = \begin{pmatrix} 4 & 10 \\ 0 & 3 \end{pmatrix}$$

The automorphism groups of f_2 and f_4 are isomorphic to $\mathrm{PSL}_2(\mathbb{F}_{11}) \times \mathbb{Z}/2\mathbb{Z}$, while the other two are isomorphic to $\mathrm{Aut}(\mathrm{PSL}_2(\mathbb{F}_{11}))$. We have seen in Example 4.19 that the automorphism group of the p-adic origami contains $\mathrm{PSL}_2(\mathbb{F}_{11}) \times \mathbb{Z}/2\mathbb{Z}$, thus the corresponding complex origami-curve is either defined by f_2 or f_4. We can presently not decide which of the two curves is the right one, and the reason for this will become clear in the next section.

5.4 Twin curves

Consider the origami shown in Figure 5.1.

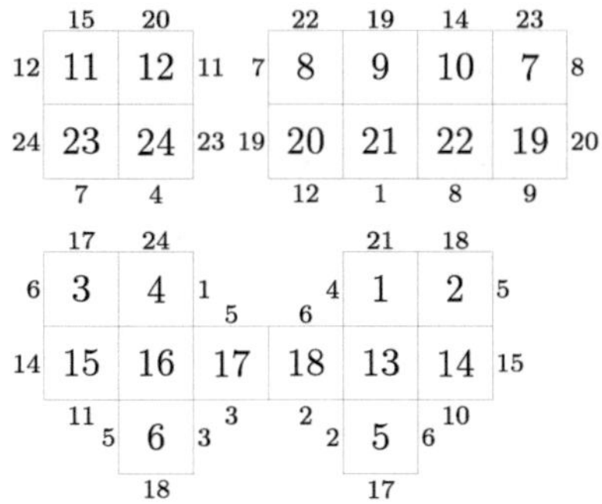

Figure 5.1: An origami O whose curve is not fixed by complex conjugation

This origami has 12 unramified punctures and four punctures with ramification index 3. Its genus is $g = 5$. The index of the (projective) Veech group is 6. The origami-curve has genus 0 and has three cusps, all of width 2. The intersection graphs of the stable curves at these cusps are shown in Figure 5.2. There is only one non-trivial automorphism, which is even a hyperellip-

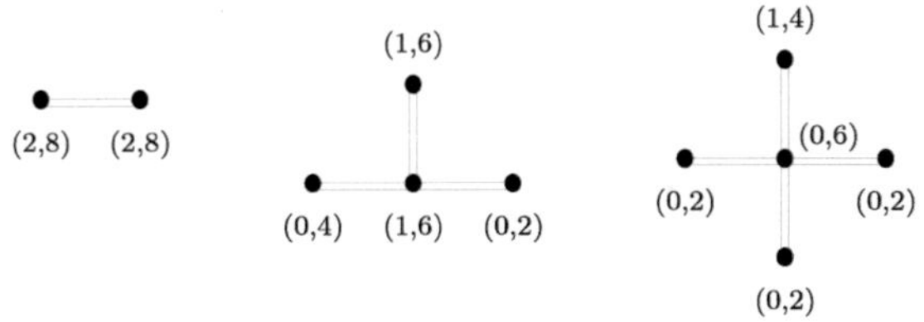

Figure 5.2: Intersection graphs of the stable curves at the cusps of $c(O)$

tic involution. It fixes two points of ramification index 3 and ten points of ramification index 0. A canonical spin structure exists, its parity is even.

There is a natural action of the absolute Galois group $\mathrm{Gal}(\overline{\mathbb{Q}}/\mathbb{Q})$ on origami-curves, and by [Möl], Theorem 5.4 this action is faithful, i.e. for every $\sigma \in \mathrm{Gal}(\overline{\mathbb{Q}}/\mathbb{Q})$ there is an origami-curve c which is mapped to another origami-curve $c^\sigma \neq c$ by σ. One would expect the properties of the curves c and c^σ to be quite similar.

The origami above (constructed by Florian Nisbach, [Nis]) is an example for an origami-curve which is not fixed by the complex conjugation, but is instead mapped to the curve of the origami shown in Figure 5.3.

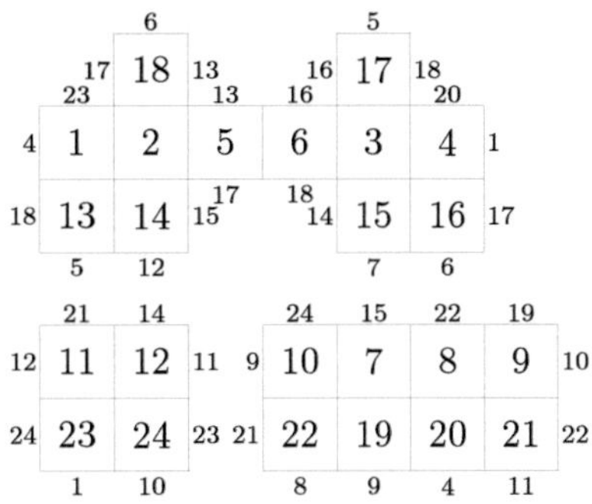

Figure 5.3: The image of the $c(O)$ under complex conjugation

Not surprisingly all the properties stated above hold for this origami as well. We therefore arrive at

Conjecture 5.8. *All the invariants of origamis introduced in chapter 3 are also invariant under the action of* $\mathrm{Gal}(\overline{\mathbb{Q}}/\mathbb{Q})$.

This would mean that we have to accept the fact that sometimes we cannot distinguish certain origami-curves just by using our invariants.

Appendix A

Algorithms and calculations

A.1 Counting origami-curves

This is an implementation of the algorithm outlined in section 2.2 written for the computer algebra system MAGMA. It calculates a representative for every origami-curve of origamis of degree d. The results up to degree 8 are listed in section A.2.

Note that as the number of vertices of our graph grows exponentially we don't explicitly construct the graph completely before computing its connected components to save memory. We rather choose a vertex, delete it, compute its neighbors, continue with deleting those, and so on, until we deleted a complete connected component. Thus the only thing we have to save is whether a vertex is deleted or not.

```
Components := function(d)
   Sd:=SymmetricGroup(d);
   Idx := SetToIndexedSet(Set(Sd));
   V := [[true : x in [1..#Sd]] : y in [1..#Sd]];
   count := #Sd^2;
   result := [];

   // Generators of Inn(Sd): conjugations with transpositions (1 i)

   GenInnSd := [hom<Sd->Sd |
               s:->Sd![{@1,i@}]*s*Sd![{@1,i@}]> : i in [2..d]];

   while count gt 0 do    // while there are vertices in the graph

      // find an unused vertex (i,j)
```

```
      i := 0;
      repeat
         i := i+1;
         j := Index(V[i],true);
      until i ge #Sd or j ne 0;

      // if the subgroup generated by i and j is transitive
      // the vertex (i,j) describes an origami: then save it

      if IsTransitive(sub<Sd | [Idx[i],Idx[j]]>) then
         Append(~result, <Idx[i],Idx[j]>);
      end if;

      // delete the connected component of (i,j)

      V[i][j] := false;
      count := count - 1;
      todo := [<i,j>];
      while not IsEmpty(todo) do
         v := todo[#todo];                // pick a vertex v
         Prune(~todo);

         // calculate the endpoints of all edges starting at v
         s := <Idx[v[1]],Idx[v[2]]>;
         E := [ <s[1],s[1]*s[2]>,         // GenOutF2: x->x, y->x*y
                <s[2],s[1]^(-1)>]         //           x->y, y->x^(-1)
              cat [<f(s[1]),f(s[2])> : f in GenInnSd];

         for e in E do                   // for all neighbours of v
            n := <Index(Idx, e[1]), Index(Idx, e[2])>;
            if V[n[1]][n[2]] then         // if they still exist
               V[n[1]][n[2]] := false;    // delete them from the graph
               count := count-1;          // and queue them for
               Append(~todo, n);          // further calculation
            end if;
         end for;
      end while;

   end while;
   return result;
end function;
```

We can do the same thing for normal origamis by selecting a Galois group G and then consider the graph with vertex set $G \times G$. The only things that change are the calculation of the generators of $\mathrm{Aut}(G)$ and the check for surjectivity instead of transitivity.

Given a degree d one can use the small groups library of MAGMA to find all groups of cardinality d, and thus one can easily create representatives of all normal origami-curves of degree d.

```
Components := function(G)
   Idx := SetToIndexedSet(Set(G));
   V := [[true : x in [1..#G]] : y in [1..#G]];
   count := #G^2;
   result := [];

   // calculate generators of Aut(G)

   p,AutG := PermutationRepresentation(AutomorphismGroup(G));
   GenAutG := [Inverse(p)(f) : f in Generators(AutG)];

   while count gt 0 do    // while there are vertices in the graph

      // find an unused vertex (i,j)

      i := 0;
      repeat
         i := i+1;
         j := Index(V[i],true);
      until i ge #G or j ne 0;

      // if the subgroup of G generated by i and j is G itself
      // the vertex (i,j) describes an origami: then save it

      if #sub<G | [Idx[i],Idx[j]]> eq #G then
         Append(~result, <Idx[i],Idx[j]>);
      end if;

      // delete the connected component of (i,j)

      V[i][j] := false;
      count := count - 1;
      todo := [<i,j>];
      while not IsEmpty(todo) do
         v := todo[#todo];                // pick a vertex v
```

```
            Prune(~todo);

            // calculate the endpoints of all edges starting at v
            s := <Idx[v[1]],Idx[v[2]]>;
            E := [ <s[1],s[1]*s[2]>,           // GenOutF2: x->x, y->x*y
                     <s[2],s[1]^(-1)>]         //           x->y, y->x^(-1)
                  cat [<f(s[1]),f(s[2])> : f in GenAutG];

            for e in E do                      // for all neighbours of v
               n := <Index(Idx, e[1]), Index(Idx, e[2])>;
               if V[n[1]][n[2]] then           // if they still exist
                  V[n[1]][n[2]] := false;      // delete them from the graph
                  count := count-1;            // and queue them for
                  Append(~todo, n);            // further calculation
               end if;
            end for;
         end while;

      end while;
      return result;
end function;
```

A.2 Origami curves up to degree 8

We use the algorithm described in section A.1 to list all curves of origamis with up to eight squares. For each of the 207 origami-curves we list a representing origami given by permutations σ_x and σ_y. We also calculate some of the invariants described in chapter 3:

d	the degree of the origami
g	the genus of the origami
ram.	the ramification indices of the punctures
$\boldsymbol{d}$	the index of the (projective) Veech group
$\boldsymbol{g}$	the genus of the origami-curve
$\boldsymbol{n}$	the number of cusps of the origami-curve
Aut	the number of automorphisms
-1	is -1 an element of the Veech group? If there is even a hyperelliptic involution we mark this with 'h'.
$\mathscr{H}$	the connected component in the moduli space $\mathscr{H}$ of Abelian differentials ('h' if the property "hyperelliptic" is satisfied, 0 or 1 for the parity of the canonical spin structure if it exists)

σ_x	σ_y	d	g	ram.	$\boldsymbol{d}$	$\boldsymbol{g}$	$\boldsymbol{n}$	Aut	-1	$\mathscr{H}$
id	id	1	1	1	1	0	1	2	h	1
(1 2)	(1 2)	2	1	1 1	3	0	2	4	h	1
id	(1 3 2)	3	1	1 1 1	4	0	2	6	h	1
(1 2)	(2 3)	3	2	3	3	0	2	2	h	h 1
(1 2)(3 4)	(1 3)(2 4)	4	1	1 1 1 1	1	0	1	8	h	1
id	(1 4 3 2)	4	1	1 1 1 1	6	0	3	8	h	1
(1 2)	(2 4 3)	4	2	1 3	9	0	3	2	h	h 1
(1 3 2)	(1 2)(3 4)	4	2	2 2	4	0	2	2	h	h
(1 2)	(1 3)(2 4)	4	2	2 2	6	0	3	4	h	h
id	(1 5 4 3 2)	5	1	1 1 1 1 1	6	0	2	10	h	1
(1 3 2)	(3 5 4)	5	2	1 1 3	9	0	3	2	h	h 1
(1 2)	(2 5 4 3)	5	2	1 1 3	18	0	5	2	h	h 1
(1 2)	(1 3)(2 5 4)	5	2	1 2 2	24	0	6	2	h	h
(1 2)(3 4)	(2 3)(4 5)	5	3	5	3	0	2	2	h	h 0
(1 2)(3 4)	(2 5 4 3)	5	3	5	6	0	3	1		1
(1 3 2)	(2 4)(3 5)	5	3	5	10	0	3	2	×	1
(1 3 2)	(1 4 2)(3 5)	5	3	5	15	0	4	2	h	h 0
id	(1 6 5 4 3 2)	6	1	1 1 1 1 1 1	12	0	4	12	h	1
(1 2)(3 4)	(1 6 5 4 2 3)	6	2	1 1 1 3	3	0	2	2	h	h 1
(1 2)(3 4)	(1 3)(2 6 5 4)	6	2	1 1 1 3	6	0	3	2	h	h 1
(1 2)	(2 6 5 4 3)	6	2	1 1 1 3	36	0	8	2	h	h 1
(1 2)(3 4)	(1 5 3)(2 6 4)	6	2	1 1 2 2	4	0	2	4	h	h
(1 2)	(1 4 3)(2 6 5)	6	2	1 1 2 2	12	0	4	4	h	h
(1 3 2)	(1 2)(3 6 5 4)	6	2	1 1 2 2	24	0	6	2	h	h
(1 4 3 2)	(1 3 2)(4 6 5)	6	2	1 1 2 2	24	0	6	2	h	h
(1 2)	(1 3)(2 6 5 4)	6	2	1 1 2 2	24	0	6	2	h	h
(1 2)(3 4)	(1 3 2)(4 6 5)	6	3	1 5	10	0	3	2	h	h 0
(1 2)(3 4)	(2 6 5 4 3)	6	3	1 5	10	0	3	1		1
(1 3 2)	(1 4 2)(3 6 5)	6	3	1 5	15	0	4	2	h	h 0
(1 2)(3 4)	(1 6 5 4 3 2)	6	3	1 5	15	0	4	2	×	1
(1 3 2)	(1 2 4)(3 6 5)	6	3	1 5	15	0	4	2	h	h 0
(1 2)(3 4)	(2 3)(4 6 5)	6	3	1 5	30	0	7	2	h	h 0
(1 3 2)	(2 4)(3 6 5)	6	3	1 5	60	0	12	1		1
(1 2)(3 4)	(1 5 3 2)(4 6)	6	3	2 4	16	0	4	1		
(1 2)(3 4)	(2 5 3)(4 6)	6	3	2 4	48	0	10	1		
(1 2)(3 4)(5 6)	(1 3)(2 5)(4 6)	6	3	3 3	3	0	2	12	h	1
(1 2)(3 4)	(1 3)(2 5)(4 6)	6	3	3 3	6	0	3	4	h	h 0
(1 3 2)	(1 5 2 4)(3 6)	6	3	3 3	6	0	3	2	h	h 0
(1 2)(3 4)	(1 6 4 5 2)	6	3	3 3	9	0	3	2	h	1
(1 2)(3 4)	(1 3)(2 6 4 5)	6	3	3 3	9	0	3	4	h	1
(1 4 3 2)	(1 3)(2 5)(4 6)	6	3	3 3	9	0	3	4	h	h 0

σ_x	σ_y	d	g	ram.	$\boldsymbol{d}$	$\boldsymbol{g}$	$\boldsymbol{n}$	Aut	-1	$\mathscr{H}$
(1 3 2)	(1 4)(2 5)(3 6)	6	3	3 3	12	0	4	6	×	1
(1 2)(3 4)	(1 6 4 5 2 3)	6	3	3 3	18	0	5	2	×	1
(1 2)(3 4)	(2 6 4 5)	6	3	3 3	18	0	5	4	h	1
(1 4 3 2)	(1 5 3)(4 6)	6	3	3 3	36	0	8	2	h	h 0
id	(1 7 6 5 4 3 2)	7	1	1 1 1 1 1 1 1	8	0	2	14	h	1
(1 3 2)	(3 7 6 5 4)	7	2	1 1 1 1 3	36	0	8	2	h	h 1
(1 2)	(2 7 6 5 4 3)	7	2	1 1 1 1 3	54	0	10	2	h	h 1
(1 2)(3 4)	(1 7 6 4 2 5 3)	7	2	1 1 1 2 2	16	0	4	2	h	h
(1 2)	(1 3)(2 7 6 5 4)	7	2	1 1 1 2 2	144	1	24	2	h	h
(1 2)(3 4)	(2 3)(4 7 6 5)	7	3	1 1 5	30	0	7	2	h	h 0
(1 3 2)	(2 5 4)(3 7 6)	7	3	1 1 5	40	0	8	2	×	1
(1 3 2)	(2 4)(3 7 6 5)	7	3	1 1 5	60	0	12	1		1
(1 3 2)	(1 5 4 2)(3 7 6)	7	3	1 1 5	105	0	18	2	h	h 0
(1 3 2)	(1 4 2)(3 7 6 5)	7	3	1 1 5	120	0	22	2	h	h 0
(1 2)(3 4)	(2 7 6 5 4 3)	7	3	1 1 5	180	1	30	1		1
(1 2)(3 4)	(2 5 3)(4 7 6)	7	3	1 2 4	16	0	4	1		
(1 2)(3 4)	(1 7 6 4 5 3 2)	7	3	1 2 4	16	0	4	1		
(1 5 4 3 2)	(1 4 3)(5 7 6)	7	3	1 2 4	48	0	10	1		
(1 5 4 3 2)	(1 3 4)(5 7 6)	7	3	1 2 4	48	0	10	1		
(1 2)(3 4)	(1 5 3 2)(4 7 6)	7	3	1 2 4	384	1	64	1		
(1 2)(3 4)	(1 7 6 4 5 2 3)	7	3	1 3 3	18	0	5	1		1
(1 2)(3 4)	(1 3)(2 5)(4 7 6)	7	3	1 3 3	36	0	8	2	h	h 0
(1 2)(3 4)	(1 7 6 4 5 2)	7	3	1 3 3	48	0	10	1		1
(1 2)(3 4)	(1 3)(2 7 6 4 5)	7	3	1 3 3	66	0	11	2	h	1
(1 3 2)	(1 4)(2 5)(3 7 6)	7	3	1 3 3	72	0	14	2	×	1
(1 3 2)	(1 5 2 4)(3 7 6)	7	3	1 3 3	180	0	32	2	h	h 0
(1 5 4 3 2)	(1 2 6 4)(5 7)	7	3	2 2 3	24	0	6	1		
(1 2)(3 4)	(2 6 3 5)(4 7)	7	3	2 2 3	24	0	6	2	h	
(1 2)(3 4)	(1 6 3)(2 5)(4 7)	7	3	2 2 3	36	0	8	2	×	
(1 2)(3 4)	(1 5 2 6 3)(4 7)	7	3	2 2 3	72	0	14	1		
(1 2)(3 4)	(1 6 3 5 2)(4 7)	7	3	2 2 3	108	0	20	2	h	
(1 2)(3 4)(5 6)	(2 3)(4 5)(6 7)	7	4	7	3	0	2	2	h	h 0
(1 3 2)(4 6 5)	(1 2 7 6 5 3 4)	7	4	7	4	0	2	1		1
(1 4 3 2)(5 6)	(1 2)(3 7 4 5)	7	4	7	7	0	2	2	h	h 0
(1 4 3 2)(5 6)	(1 7 4 5)(2 3)	7	4	7	7	0	2	2	h	h 0
(1 2)(3 4)(5 6)	(2 3 7 6 5 4)	7	4	7	12	0	4	1		0
(1 5 4 3 2)	(1 4)(2 6 3)(5 7)	7	4	7	21	0	5	2	h	h 0
(1 5 4 3 2)	(1 6 4)(2 3)(5 7)	7	4	7	21	0	5	2	h	h 0
(1 5 4 3 2)	(1 6 4 3)(5 7)	7	4	7	28	0	6	1		0
(1 5 4 3 2)	(2 3 6 4)(5 7)	7	4	7	42	0	9	1		1
(1 5 4 3 2)	(1 2)(3 6 4)(5 7)	7	4	7	42	0	9	1		1

σ_x	σ_y	d	g	ram.	$\boldsymbol{d}$	$\boldsymbol{g}$	$\boldsymbol{n}$	Aut	-1	$\mathscr{H}$
(1432)	(163)(25)(47)	7	4	7	84	0	16	2	h	h 0
(132)(45)	(24)(3756)	7	4	7	168	0	30	1		1
(1432)	(2635)(47)	7	4	7	168	0	30	1		0
(1432)	(25)(36)(47)	7	4	7	189	0	33	2	×	1
(12)(34)(56)(78)	(1753)(2864)	8	1	11111111	3	0	2	16	h	1
id	(18765432)	8	1	11111111	12	0	4	16	h	1
(12)(34)	(13)(287654)	8	2	111113	9	0	3	2	h	h 1
(12)(34)	(18765423)	8	2	111113	18	0	5	2	h	h 1
(12)	(2876543)	8	2	111113	108	1	17	2	h	h 1
(12)(34)	(18742653)	8	2	111122	6	0	3	4	h	h
(12)(34)(56)(78)	(13)(2754)(68)	8	2	111122	6	0	3	4	h	h
(12)(34)	(1653)(2874)	8	2	111122	6	0	3	4	h	h
(132)(465)	(1524)(3876)	8	2	111122	12	0	4	2	h	h
(12)	(1543)(2876)	8	2	111122	24	0	6	4	h	h
(12)	(13)(287654)	8	2	111122	72	0	14	2	h	h
(132)	(12)(387654)	8	2	111122	96	1	16	2	h	h
(12)	(143)(28765)	8	2	111122	144	1	24	2	h	h
(12)(34)	(18765432)	8	3	1115	60	0	12	2	×	1
(1432)	(16532)(487)	8	3	1115	75	0	13	2	h	h 0
(12)(34)	(2876543)	8	3	1115	130	0	23	1		1
(12)(34)	(23)(48765)	8	3	1115	240	0	42	2	h	h 0
(132)	(142)(38765)	8	3	1115	270	1	45	2	h	h 0
(132)	(24)(38765)	8	3	1115	510	5	77	1		1
(12)(34)	(2876453)	8	3	1124	48	0	10	1		
(12)(34)	(1532)(4876)	8	3	1124	48	0	10	1		
(132)(45)	(14)(263875)	8	3	1124	96	0	18	1		
(12)(34)	(16532)(487)	8	3	1124	384	1	64	1		
(12)(34)	(253)(4876)	8	3	1124	1440	16	210	1		
(132)(465)	(1425)(3876)	8	3	1133	3	0	2	2	h	1
(132)(465)	(14)(25)(37)(68)	8	3	1133	9	0	3	4	h	h 0
(132)(465)	(187635)(24)	8	3	1133	12	0	4	1		1
(12)(34)	(1523)(4876)	8	3	1133	18	0	5	2	h	h 0
(1432)(56)	(15)(26)(37)(48)	8	3	1133	18	0	5	2	×	1
(12)(34)	(13)(287465)	8	3	1133	18	0	5	4	h	1
(132)(465)	(354)(687)	8	3	1133	18	0	5	2	h	h 0
(1432)	(265)(487)	8	3	1133	36	0	7	4	h	1
(132)	(1524)(3876)	8	3	1133	36	0	8	2	h	h 0
(12)(34)(56)	(18674253)	8	3	1133	36	0	8	2	×	1
(12)(34)	(13)(25)(4876)	8	3	1133	36	0	8	2	h	h 0
(12)(34)	(13)(265)(487)	8	3	1133	36	0	8	4	h	h 0
(12)(34)	(287465)	8	3	1133	36	0	8	4	h	1

σ_x	σ_y	d	g	ram.	$\boldsymbol{d}$	$\boldsymbol{g}$	$\boldsymbol{n}$	Aut	-1	$\mathscr{H}$
(1 2)(3 4)	(1 8 7 6 4 5 2)	8	3	1 1 3 3	36	0	8	1		1
(1 2)(3 4)(5 6)	(1 8 6 7 4 2 3)	8	3	1 1 3 3	45	0	9	2	h	1
(1 3 2)	(1 4)(2 5)(3 8 7 6)	8	3	1 1 3 3	54	0	11	2	×	1
(1 4 3 2)	(2 5)(4 8 7 6)	8	3	1 1 3 3	60	0	12	2	h	1
(1 2)(3 4)	(1 3)(2 8 7 6 4 5)	8	3	1 1 3 3	90	1	15	2	h	1
(1 4 3 2)	(1 6 5 3)(4 8 7)	8	3	1 1 3 3	144	0	26	2	h	h 0
(1 3 2)	(1 4)(2 6 5)(3 8 7)	8	3	1 1 3 3	144	1	24	2	×	1
(1 3 2)	(1 6 5 2 4)(3 8 7)	8	3	1 1 3 3	198	0	35	2	h	h 0
(1 2)(3 4)	(1 8 7 6 4 5 2 3)	8	3	1 1 3 3	312	3	48	1		1
(1 4 3 2)	(1 5 3)(4 8 7 6)	8	3	1 1 3 3	378	4	57	2	h	h 0
(1 3 2)(4 6 5)	(1 5)(2 4)(3 8 7 6)	8	3	1 2 2 3	6	0	3	2	h	
(1 2)(3 4)	(1 5 2 8 7 4 6 3)	8	3	1 2 2 3	18	0	5	2	×	
(1 2)(3 4)	(1 8 7 4 6 3)(2 5)	8	3	1 2 2 3	18	0	5	1		
(1 4 3 2)(5 6)	(1 5)(2 7 3 6)(4 8)	8	3	1 2 2 3	24	0	6	1		
(1 2)(3 4)	(1 8 7 4 6 2 3 5)	8	3	1 2 2 3	36	0	8	2	×	
(1 2)(3 4)	(2 6 3 5)(4 8 7)	8	3	1 2 2 3	192	0	34	2	h	
(1 2)(3 4)	(1 5 2 6 3)(4 8 7)	8	3	1 2 2 3	216	1	36	1		
(1 2)(3 4)	(1 6 3 5 2)(4 8 7)	8	3	1 2 2 3	252	1	42	2	h	
(1 2)(3 4)	(1 6 3)(2 5)(4 8 7)	8	3	1 2 2 3	840	10	122	1		
(1 4 3 2)(5 8 7 6)	(1 5 3 7)(2 8 4 6)	8	3	2 2 2 2	1	0	1	16	×	
(1 2)(3 4)(5 6)(7 8)	(1 3)(2 5)(4 7)(6 8)	8	3	2 2 2 2	3	0	2	16	h	
(1 2)(3 4)	(1 8 4 6 2 7 3 5)	8	3	2 2 2 2	6	0	3	8	×	
(1 2)(3 4)	(1 7 3 5)(2 8 4 6)	8	3	2 2 2 2	6	0	3	8	h	
(1 7 6 5 4 3 2)	(1 5)(2 3)(4 6)(7 8)	8	3	2 2 2 2	8	0	2	1		
(1 2)(3 4)	(1 7 3 6 2 5)(4 8)	8	3	2 2 2 2	12	0	4	2	h	
(1 4 3 2)(5 6)	(1 5)(2 7)(3 6)(4 8)	8	3	2 2 2 2	12	0	4	4	h	
(1 3 2)(4 6 5)	(1 5)(2 4)(3 7)(6 8)	8	3	2 2 2 2	12	0	4	4	h	
(1 2)(3 4)	(1 5)(2 7 3 6)(4 8)	8	3	2 2 2 2	24	0	6	4	h	
(1 2)(3 4)(5 6)	(2 3 8 6 4 7 5)	8	3	2 2 2 2	36	0	7	2	×	
(1 3 2)(4 6 5)	(1 2)(3 8 6 7)(4 5)	8	3	2 2 2 2	48	0	10	4	h	
(1 3 2)(4 5)	(1 2)(3 7 4 6)(5 8)	8	3	2 2 2 2	132	1	19	2	h	
(1 3 2)(4 6 5)	(1 7 5 3)(2 8 6 4)	8	4	1 7	7	0	2	2	×	0
(1 4 3 2)(5 6)	(1 5 3 2)(4 8 6 7)	8	4	1 7	21	0	5	2	×	1
(1 4 3 2)(5 6)	(1 5 3)(4 8 7)	8	4	1 7	21	0	5	2	h	h 0
(1 4 3 2)(5 6)	(1 2 3 5)(4 8 6 7)	8	4	1 7	21	0	5	2	×	1
(1 4 3 2)(5 6)	(1 3 5)(4 8 7)	8	4	1 7	21	0	5	2	h	h 0
(1 3 2)(4 6 5)	(2 4 3)(6 8 7)	8	4	1 7	28	0	6	1		0
(1 3 2)(4 6 5)	(1 4 3)(6 8 7)	8	4	1 7	28	0	6	1		0
(1 5 4 3 2)	(1 7 4 2 6 3)(5 8)	8	4	1 7	42	0	9	2	h	h 0
(1 2)(3 4)(5 6)	(2 5 4)(6 8 7)	8	4	1 7	42	0	9	2	×	1
(1 2)(3 4)(5 6)	(2 3)(4 5)(6 8 7)	8	4	1 7	42	0	9	2	h	h 0

σ_x	σ_y	d	g	ram.	$\boldsymbol{d}$	$\boldsymbol{g}$	$\boldsymbol{n}$	Aut	-1	$\mathscr{H}$
(1432)	(15263)(487)	8	4	1 7	336	1	56	2	×	1
(15432)	(12643)(587)	8	4	1 7	546	1	91	1		0
(15432)	(12)(46)(587)	8	4	1 7	546	2	89	1		1
(1432)	(163)(25)(487)	8	4	1 7	567	5	85	2	h	h 0
(1432)	(2635)(487)	8	4	1 7	1260	15	182	1		0
(1432)	(25)(36)(487)	8	4	1 7	1932	22	280	1		1
(1432)(56)	(275)(486)	8	4	2 6	18	0	5	2		
(12)(34)(56)	(275)(486)	8	4	2 6	18	0	5	2		
(1432)(56)	(186523)(47)	8	4	2 6	36	0	8	1		
(12)(34)(56)	(13)(275)(486)	8	4	2 6	36	0	8	2		
(1432)(56)	(138652)(47)	8	4	2 6	36	0	8	1		
(15432)	(263)(47)(58)	8	4	2 6	648	1	108	1		
(132)(45)	(26374)(58)	8	4	2 6	2160	22	318	1		
(1432)(56)	(25)(4867)	8	4	3 5	15	0	4	2	×	1
(12)(34)(56)	(23)(4867)	8	4	3 5	30	0	7	2	h	0
(15432)	(174362)(58)	8	4	3 5	45	0	9	2	h	0
(15432)	(26)(47)(58)	8	4	3 5	180	0	32	2	×	1
(1432)(56)	(12)(4867)	8	4	3 5	240	1	40	1		1
(15432)	(127463)(58)	8	4	3 5	270	0	47	1		0
(132)(45)	(162)(3857)	8	4	3 5	330	3	50	2	h	0
(132)(45)	(173854)(26)	8	4	3 5	540	5	82	1		0
(132)(45)	(26)(3857)	8	4	3 5	900	11	130	1		1
(12)(34)(56)	(13)(25)(47)(68)	8	4	4 4	6	0	3	4	h	h
(12)(34)(56)	(1523)(4867)	8	4	4 4	6	0	3	4	×	
(1432)	(1735)(26)(48)	8	4	4 4	12	0	4	4	h	h
(12)(34)(56)	(13)(25)(4867)	8	4	4 4	12	0	4	4	×	
(132)(465)	(12)(34)(57)(68)	8	4	4 4	16	0	4	2	×	
(132)(465)	(1542)(3867)	8	4	4 4	16	0	4	4	×	
(132)(465)	(1254)(3867)	8	4	4 4	16	0	4	2	×	
(12)(34)(56)	(18674352)	8	4	4 4	24	0	6	2	×	
(12)(34)(56)	(286745)	8	4	4 4	24	0	6	2	×	
(1432)	(15)(26)(37)(48)	8	4	4 4	24	0	6	8	×	
(132)(45)	(1624)(37)(58)	8	4	4 4	24	0	6	2	h	h
(165432)	(12)(37)(45)(68)	8	4	4 4	24	0	6	4	×	
(132)(465)	(1542)(37)(68)	8	4	4 4	24	0	6	4	h	h
(1432)(56)	(12)(35)(47)(68)	8	4	4 4	48	0	10	1		
(1432)(56)	(2374865)	8	4	4 4	48	0	10	1		
(1432)	(1625)(37)(48)	8	4	4 4	72	0	14	2	×	
(1432)(56)	(153)(47)(68)	8	4	4 4	96	0	18	2	h	h
(132)(45)	(1624)(3857)	8	4	4 4	240	0	42	2	×	
(15432)	(174)(263)(58)	8	4	4 4	288	0	50	2	h	h

σ_x	σ_y	d	g	ram.	$\boldsymbol{d}$	$\boldsymbol{g}$	$\boldsymbol{n}$	Aut	-1	$\mathscr{H}$
(1 5 4 3 2)	(1 2)(3 6)(4 7)(5 8)	8	4	4 4	384	2	62	2	×	
(1 3 2)(4 5)	(1 4)(2 6)(3 8 5 7)	8	4	4 4	576	6	86	1		

A.3 Groups whose curves are not unique

There are 6960 groups whose order is less than or equal to 250. Of these 2386 can be generated by two elements and are thus possible Galois groups of origamis. Almost all of the corresponding origami-curves are already uniquely defined by their Galois group. There are only 30 exceptions, which are listed in the following table. In the columns we have noted first the group order n, then the number k of the group in the small-groups-library of MAGMA (i.e. the group can be constructed in MAGMA using the command `SmallGroup(n,k)`) and finally the number of origami-curves and their ramification indices:

order	number	curves	ramification
60	5	2	3, 5
81	10	2	3, 3
120	5	2	6, 10
120	34	3	2, 3, 5
120	35	2	3, 5
160	199	2	2, 4
162	31	2	3, 3
168	42	4	3, 4, 4, 7
168	43	2	7, 7
180	19	2	3, 5
189	7	2	21, 21
192	181	2	6, 6
192	201	3	2, 4, 4
192	202	2	2, 4
192	1491	2	6, 6
200	44	3	2, 2, 2
216	87	2	2, 6
216	153	3	4, 4, 4
240	89	3	3, 4, 10
240	90	3	3, 4, 10
240	91	3	2, 3, 5
240	92	2	3, 5
240	93	2	6, 10
240	94	2	6, 10
240	189	3	2, 3, 5

order	number	curves	ramification
240	190	2	3, 5
243	5	2	3, 3
243	7	2	3, 3
243	16	2	3, 3
243	18	2	3, 3

A.4 Determining the automorphism groups

The following MAGMA code calculates the automorphism group of a normal origami given by $f : F_2 \twoheadrightarrow G$, more precisely by the images $\sigma := f(x)$ and $\tau := f(y)$ of the generators. We have seen in Proposition 3.13 that this group is isomorphic to $C_2 \ltimes_\Phi G$ where $\Phi : C_2 \to \mathrm{Aut}(G)$ maps the generator of the cyclic group C_2 to $\varphi \in \mathrm{Aut}(G)$ defined by $\varphi(\sigma) = \sigma^{-1}$ and $\varphi(\tau) = \tau^{-1}$. If such an automorphism doesn't exist the automorphism group equals the Galois group.

```
Aut := function(sigma,tau)
   G := sub<Parent(sigma)|[sigma,tau]>;
   A := AutomorphismGroup(G);
   if not IsHomomorphism(G,G,[Inverse(sigma), Inverse(tau)])
      then
         return G;
      else
         phi := hom<G -> G | [Inverse(sigma), Inverse(tau)]>;
         C2 := CyclicGroup(2);
         Phi := hom<C2 -> A | [A!phi]>;
         return SemidirectProduct(G,C2,Phi);
   end if;
end function;
```

For the construction of the semidirect product we need the help of the following function, which constructs a group H from the set of its elements S and a map $\psi : S \times S \to S$ which defines the multiplication in H. This is done by identifying $y \in S$ with the map $\psi(\,\cdot\,, y) \in \mathrm{Sym}(S)$. We have

$$\psi(\,\cdot\,,x) \cdot \psi(\,\cdot\,,y) = \psi(\psi(\,\cdot\,,x),y) = \psi(\,\cdot\,,\psi(x,y))$$

where the left equation is the way MAGMA multiplies permutations and the right equation holds because ψ has to be associative. Thus we get a subgroup H of $\mathrm{Sym}(S)$ which respects the group operation defined by ψ.

```
Group := function(S, psi)
   G := Sym(S);
   H := sub<G | [G![psi(<x,y>) : x in S] : y in S]>;
   BSGS(H);
   ReduceGenerators(~H);
   E := {x : x in S | forall{y : y in S | psi(<x,y>) eq y}};
   e := Rep(E);  // the neutral Element
   f := map<S->H | y :-> H![psi(<x,y>): x in S], p :-> Image(p,S!e)>;
   return H,f;
end function;
```

We can now construct a semidirect product $G := H \ltimes_{\phi} N$ with $\varphi : H \to N$ by using the multiplication on $S := N \times H$ defined by $\psi : S \times S \to S$ with $\psi((n_1, h_1), (n_2, h_2)) = (n_1\varphi(h_1)(n_2), h_1h_2)$. The function also returns the inclusions $i_N : N \hookrightarrow G$, $i_H : H \hookrightarrow G$ and the projection $p_H : G \twoheadrightarrow H$.

```
SemidirectProduct := function(N,H,phi)
   S := Set(CartesianProduct(Set(N),Set(H)));
   S2 := CartesianProduct(S,S);
   psi := map<S2->S | T :-> S!<N!T[1][1]*phi(H!T[1][2])(N!T[2][1]),
                                 H!T[1][2]*H!T[2][2]> >;
   G,f := Group(S,psi);
   iN := hom<N -> G | x :-> f(<x,H!1>) >;
   iH := hom<H -> G | x :-> f(<N!1,x>) >;
   pH := hom<G -> H | g :-> (g @@ f)[2]>;
   if #N*#H eq #G
      then return G,iN,pH,iH;
      else return false;
   end if;
end function;
```

Index

Bibliography

[Ati] M. F. ATIYAH: Riemann Surfaces and Spin Structures. Annales scientifiques de l'École Normale Supérieure, Série 4, Vol. 4, No. 1, pp. 47-62 (1971)

[Bra1] P. E. BRADLEY: Mumfordkurven und Reduktion analytischer Abbildungen. Diplomarbeit, Karlsruhe (1998)

[Bra2] P. E. BRADLEY: Riemann Existence Theorems of Mumford Type. Mathematische Zeitschrift 251 No. 2, pp. 393-414 (2005)

[Bra3] P. E. BRADLEY: Cyclic coverings of the p-adic projective line by Mumford curves. Manuscripta Mathematica 124, pp. 77-95 (2007)

[BKV] P. E. BRADLEY, F. KATO, H. VOSKUIL: The Classification of p-adic Discrete Subgroups of PGL_2 *(unpublished)*

[DM] P. DELIGNE, D. MUMFORD: The irreducibility of the space of curves of given genus. Publications Mathématiques d l'IHÉS 36, pp. 75-110 (1969)

[EO] A. ESKIN, A. OKOUNKOV: Asymptotics of numbers of branched coverings of a torus and volumes of moduli spaces of holomorphic differentials. Inventiones mathematicae 145, pp. 59-103 (2001)

[FP] J. FRESNEL, M. VAN DER PUT: Rigid Analytic Geometry and its Applications. Progress in Mathematics 218, Birkhäuser (2003)

[GJ] E. GUTKIN, C. JUDGE: Affine mappings of translation surfaces. Duke Mathematical Journal 103 No. 2, pp.191-212 (2000)

[GP] L. GERRITZEN, M. VAN DER PUT: Schottky groups and Mumford curves. Lecture Notes in Mathematics 817, Springer (1980)

[Har] R. HARTSHORNE: Algebraic Geometry. Graduate Texts in Mathematics 52, Springer (1977)

[Her] F. HERRLICH: p-adisch diskontinuierlich einbettbare Graphen von Gruppen. Archiv der Mathematik 39 No. 3, pp. 204-216 (1982)

[HL] P. HUBERT, S. LELIÈVRE: Prime arithmetic Teichmüller discs in $\mathcal{H}(2)$. Israel Journal of Mathematics 151, pp. 281-322 (2006)

[HS1] F. HERRLICH, G. SCHMITHÜSEN: A comb of origami curves in the moduli space M_3 with three dimensional closure. Geometriae dedicata 124, pp. 69-94 (2007).

[HS2] F. HERRLICH, G. SCHMITHÜSEN: On the boundary of Teichmüller disks in Teichmüller and in Schottky space. Handbook of Teichmüller Theory Vol. I (ed. A. Papadopoulos) EMS, pp. 293-349 (2007)

[HS3] P. HUBERT, T. A. SCHMIDT: An introduction to Veech surfaces. Handbook of Dynamical Systems 1B, pp. 501-526 (2006)

[JR] G. JANUSZ, J. ROTMAN: Outer Automorphisms of S_6. The American Mathematical Monthly 89 No. 6, pp. 407-410 (1982)

[Kap] A. KAPPES: On the Equation of an Origami of Genus two with two Cusps. Diplomarbeit, Karlsruhe (2007)

[Kat1] F. KATO: p-adic Schwarzian triangle groups of Mumford type. Preprint, arXiv:math/9908174v3 (2000)

[Kat2] F. KATO: Non-archimedean orbifolds covered by Mumford curves. Journal of Algebraic Geometry 14, pp. 1-34 (2005)

[Kre] K. KREMER: Galoissche Mumford-Origamis. Diplomarbeit, Karlsruhe (2005)

[KZ] M. KONTSEVICH, A. ZORICH: Connected components of the moduli spaces of Abelian differentials with prescribed singularities. Inventiones mathematicae 153, pp. 631-678 (2003)

[Lef] S. LEFSCHETZ: Algebraic Geometry. Princeton University Press (1953)

[Loc] P. LOCHAK: On Arithmetic Curves in the Moduli Space of Curves. Journal of the Institute of Mathematics of Jussieu 4 No. 3, pp. 443-508 (2005)

[LS] R. C. LYNDON, P. E. SCHUPP: Combinatorial Group Theory. Springer (1977)

[Mc1] C. T. McMullen: Billiards and Teichmüller curves on Hilbert surfaces. Journal of the American Mathematical Society 16 No. 4, pp. 857-885 (2003)

[Mc2] C. T. McMullen: Teichmüller curves in genus two: Discriminant and spin. Mathematische Annalen 333, pp. 87-130 (2005)

[Möl] M. Möller: Teichmüller curves, Galois actions and GT-relations. Mathematische Nachrichten 278 No. 9, pp. 1061-1077 (2005)

[Mum] A. Mumford: An analytic construction of degenerating curves over complete local rings. Compositio Mathematica 24, pp. 129-174 (1972)

[Nag] S. Nag: The Complex Analytic Theory of Teichmüller Spaces. Wiley (1988)

[Nis] F. Nisbach: Dissertation, Karlsruhe *(to appear)*

[PV] M. van der Put, H. Voskuil: Mumford Coverings of the projective line. Archiv der Mathematik 80, pp. 98-105 (2003)

[Rot] J. Rotman: The Theory of Groups, An Introduction, 2nd ed. Allyn & Bacon, Boston, MA (1973)

[Sch1] G. Schmithüsen: Veech Groups of Origamis. Dissertation, Karlsruhe (2005)

[Sch2] G. Schmithüsen: An algorithm for finding the Veech group of an origami. Experimental Mathematics 13, pp. 459-472 (2004)

[Ser] J.-P. Serre: Trees. Springer (1980)

[Sti] J. Stillwell: Classical Topology and Combinatorial Group Theory. Graduate Texts in Mathematics 72, Springer (1980)

[Thu] W. Thurston: On the geometry and dynamics of diffeomorphisms of surfaces. Bulletin of the American Mathematical Society 19 No. 2, pp. 417-431 (1988)

[Vee] W. A. Veech: Teichmüller curves in moduli space, Eisenstein series and an application to triangular billards. Inventiones Mathematicae 97 No. 3, pp. 553-583 (1989)

[Zor] A. Zorich: Square Tiled Surfaces and Teichmüller Volumes of the Moduli Spaces of Abelian Differentials.
in: M. Burger, A. Iozzi (Ed.): Rigidity in Dynamics and Geometry, pp. 459-471, Springer (2002)